FOCUS FOR EXCELLENCE
LABORATORY PRACTICE

FOCUS FOR EXCELLENCE
LABORATORY PRACTICE

VIROJ WIWANITKIT

Nova Biomedical Books
New York

For permission to use material from this book please contact us:
Telephone 631-231-7269; Fax 631-231-8175
Web Site: http://www.novapublishers.com

NOTICE TO THE READER

The Publisher has taken reasonable care in the preparation of this book, but makes no expressed or implied warranty of any kind and assumes no responsibility for any errors or omissions. No liability is assumed for incidental or consequential damages in connection with or arising out of information contained in this book. The Publisher shall not be liable for any special, consequential, or exemplary damages resulting, in whole or in part, from the readers' use of, or reliance upon, this material.

Independent verification should be sought for any data, advice or recommendations contained in this book. In addition, no responsibility is assumed by the publisher for any injury and/or damage to persons or property arising from any methods, products, instructions, ideas or otherwise contained in this publication.

This publication is designed to provide accurate and authoritative information with regard to the subject matter covered herein. It is sold with the clear understanding that the Publisher is not engaged in rendering legal or any other professional services. If legal or any other expert assistance is required, the services of a competent person should be sought. FROM A DECLARATION OF PARTICIPANTS JOINTLY ADOPTED BY A COMMITTEE OF THE AMERICAN BAR ASSOCIATION AND A COMMITTEE OF PUBLISHERS.

Library of Congress Cataloging-in-Publication Data
Viroj Wiwanitkit.
  Focus for excellence laboratory practice / Viroj Wiwanitkit.
    p. cm.
  Includes index.
  ISBN 978-1-60692-585-0 (softcover)
  1. Diagnosis, Laboratory. I. Title.
  RB37.V57   2009
  616.07'56--dc22   2008051253

Published by Nova Science Publishers, Inc. ✛ *New York*

Contents

Preface

Laboratory Investigation is a big tool for biological and medical work around the world. The present concept of laboratory investigation is to provide the best analysis. This covers several facets and all phases of the laboratory cycle. In this book, concept and scenerios on excellent laboratory practice will be discussed and presented.

Chapter I

Quality Process for Laboratory Analysis: Introduction

Abstract

Quality is the thing that all medical laboratories need. Quality is an indicator of succeed in laboratory service. Quality in medicine means good services to users. In this article, the author will details, gives examples and discusses on quality process for laboratory analysis.

Keywords: quality, laboratory.

What Is Quality in Laboratory Medicine?

Quality is the thing that all medical laboratories need. Quality is an indicator of succeed in laboratory service. Basically, quality means everything in the process is controlled and runs in its good way. In general sense, quality can be reflected by satisfaction. In medical laboratory science, the quality must cover all phases. Monitoring of quality is also necessary. Quality in medicine means good services to users. Based on customer theory, the two main groups of users of medical laboratory are patients as external users and medical personnel in patient care unit such as wards and emergency room (doctors and nurses) as internal users. The quality management within medical laboratory should focus on these two groups of users.

There are many quality management systems at present. This can be selected as appropriate for each laboratory setting. However, the basic principle is similar. The best service is the aim of any quality management system. In order to reach the quality, the medical laboratory must sets its quality management system and uses at properly. In addition, monitoring of quality via quality indicator should be and must be done. Several indices include satisfaction, complaint, non conformation record and etc. Of several indicators, satisfaction seems to be the most important. All laboratories must set the system for the satisfaction survey of their users [1]. In addition to satisfaction survey, complaint monitoring is as important [1]. Because the uses will give the complaint to the service only if they exposed to the worst situation of service, therefore, any complaints can be good indices for the downing of service. It is necessary to accept that medical service is a kind of service commerce. Due to the fact that medical service is a kind of medial service, therefore, laboratory services also falls into the rules of service commerce. The satisfaction of the users must be the heart of this type of work. To response to minded desire, it is necessary to reach the level of satisfaction.

Quality Cycle in Laboratory Medicine

In order to understand the basic concept in laboratory medicine, the beginner must realize the basic concept of laboratory quality cycle. The laboratory quality cycle means the connection of three phases within process of whole laboratory analysis. Preanalytical phase is the first phase. This is the phase before the analytical work within the laboratory. This phase include specimen collection, transportation and preparation for further analysis. The second phase is the analytical phase. This phase is the actual practice of analysis by either medical personnel in laboratory or the modern automated analyzers. The last phase is the postanalytical phase. This phase is the process after complete analysis. This can be from validation of result to result reporting to the physicians at wards.

These three phases have to be clarified for all laboratory analysis. Deductive approach can be used to understand the factors associated with every laboratory investigation. Quality management in laboratory medicine must cover all three phases within the laboratory quality cycle as already mentioned. It should be stated that quality cycle is the stepwise process. Any situations in affection the prior phase(s) can be affect the next phase(s). Therefore, rooted cause analysis of any problematic cases in laboratory medicine must concern all steps as the whole approach. Tracing back on the step of laboratory cycle can help identify the problem in case of erroneous results. For example, if there is an error in specimen collection, it can expect for the erroneous analytical result despite good analytical and postanalytical processes. Simply, the model of input-process-output can be used. Input hereby means preanalytical phase, process means analytical phase and output means postanalytical phase. Quality management for all parts is needed.

Quality Control and Assurance

Quality control and assurance are the two important things in medical laboratory. This is the main core of classical laboratory medicine. It focuses on the analytical phase. Internal quality control, within laboratory and external quality assessment among laboratories are routinely practiced. There are the needs for internal quality control and external quality assessment for all laboratory tests within the laboratory. However, in some cases, there are limitations of availability of external control program. This can be a challenge. The author hereby presents the situation and its solution. The example on "Strategies finding for interlaboratory comparison for reticulocyte count between King Chulalongkorn Memorial Hospital and Siriraj Hospital" will be hereby presented.

1. Introduction

External quality assessment (EQA) is a requirement according to medical laboratory standards [2 - 5]. It is a point to be accredited clinical auditing [6 – 8]. An important problem in EQA management in laboratory is finding of EQA program for all available tests. Sometimes, there is no or hardly available program for some specific tests. In those cases, interlaboraoty comparison program can be useful [9]. In this work the anthers report the success in strategies finding for interlaboratory comparison for reticulocyte count between King Chulalongkorn Memorial Hospital (KCMH) and Siriraj Hospital (SH).

2. Materials and Methods

This is a kind of policy research. First, the KCMH laboratory was communicated by SH laboratory to discuss about the possibility in setting of a interlaboratory comparison program for reticulocyte count. Both laboratories are the two largest laboratories of Thailand. Both ones are certified by ISO 15189. in planning strategies, the representatives of the two laboratories were discussed in group focusing on present situation, planning for agreement and setting for communication system.

3. Results

A. Identification of Present Status of Reticulocyte
Count in Each Laboratory
KCMH laboratory uses automated analyzer namely Advia which SH laboratory uses automated analyzer namely System for reticulocyte count analysis. Both laboratories used EDTA blood samples for reticulocyte counts.

B. Planning for Agreement
Both laboratories plan for agreement based on the 95 % confidence interval of results of analyzer of EQA samples. In cases that disagreement occur, third party will be used.

C. Setting for Communication System
Both laboratories agree to perform 3 EQA rounds per years. Two samples (normal and abnormal EQA) per time was set. The messenger is planned for transporting of EQA samples between the two laboratories in appointed date. The analysis is planned performed when two laboratories get EQA samples and confirm by telephone.

4. Discussion

Interlaboratory comparison program seems to be a new program for medical laboratory in Thailand. Because the one only a few routine EQA programs and same laboratory tests lack no routine EQA program. However, with force by accreditation system, interlaboratory comparison is a new solution for these tests lacking for routine EQA program [9]. In this wake the two largest laboratories in Thailand planned for strategies in collaboration in interlaboratory comparison for reticulocyte count. Group discussion as the way of action participatory researching is done. Policy was determined by strategic planning. Both laboratories felt glad to collaborate in this program fun the status of the present technique of reticulocyte count in boll laboratories was evaluated. It was fund that different automated analyzers were used. However, this is not the contraindication for setting of interlaboratory comparison program. A different point is to set the level of agreement. The two laboratories applied the concept of 95 % confidence interval (mean + 2 Standard error) to set the agreement range of the EQA results. Also, third panty analysis was also planned for the problematic cases.

In addition the tow laboratories tired to solve the problem of communication. An indentified first is that the value of reticulocyte count decrease of prolonged collected. To avoid the time effect, the two laboratories set the transportation system via motorcycle to solver the traffic problem and making agreement confirmation system by telephoning to reassure that both laboratories will analyzer EQA samples at the same time. This report is a good example of policy research in laboratory medicine. A similar interlaboratory program is in the future plan of KCMH and SH laboratories to increase the qualities in sever of belt laboratories.

Example on Report on Result of Counseling Unit Service for Laboratory Medicine

Because the laboratory service, similar to other kinds of medical services, is not the routine knowledge for the users. Often, the patients do not know how to do. This brings the concept that there must be the information center to them at least. Here, the author also proposes the additional concept to give them the knowledge on the laboratory investigation. The situation as a model will be hereby presented and discussed.

1. Introduction

In clinical pathology, counseling is an important action. A well-known process is counseling before specimen collection in that persues Anti HIV serology test. However, there are also other types of counseling. An important counseling type is scientific counseling. This is a system of consultant in medicine.

Scientific counseling in laboratory medicine can be a useful process. This can help general physician solve specific problem in clinical pathology. In KCMH, the scientific counseling unit of Division of Laboratory Medicine has recently been serviced since June 2007. In this work, the other report the result of counseling unit service in the first month.

2. Materials and Methods

This work is a retrospective study. The record of consultations of counseling unit of Division of Laboratory Medicine, KCMH during June 2007 was reviewed. The counseling unit located at Por Por Ror Building, 4th floor and the service times one 4 hours/week. The summary of all consultant was performed by qualitative research technique.

3. Result

There was only, incidence in the student 1-month period. The consultation is about the usage of proper anticoagulant for special laboratory test, HIV-RNA. The explanation was given and the consult was completely successful.

4. Discussion

Scientific counseling can help manage the problematic case laboratory in medicine. Marques et al reported that clinical pathology consultation improved coagulation factor utilization in hospitalized adults [10]. Kondoh and Kanno also reported that the consultation could provide quick turnaround service in clinical laboratories because error could be reduced [11]. Here, the authors reported the experience of the first stage of the counseling unit for clinical pathology. The unit is a new setting aiming help the physician in the hospital for the problem on clinical pathology and also fulfill the requirement for practice in laboratory medicine [12].

Of interest, there is only one counseling. This might imply that the unit was not yet familiar to general physician in the hospital. Of interest, the basic problem in clinical pathology is not complicated and can be solved by general practitioner. However, there are also some complicated problems. In this work, the consulted problem, selection of proper additive is found. Indeed, the selection of the proper additive is hand and it seems that there is a need to generate the knowledge on this area to general physician. The results are similar to the previous report in the analysis of Yanai. Yanai said that most physicians who consult the clinical pathology counseling unit had the problem on specimen preparation, especially for microbiology test [13]. The authors hope to get here consultations in the further period and think that the service will be useful.

Summary of Concept

Laboratory investigation is a big tool for biological and medical work around the world. The present concept of laboratory investigation is to provide the best analysis. These covers several facets covering all phase of laboratory cycle. In this book, concept and scenarios on excellent laboratory practice will be discussed and presented.

References

[1]   Wiwanitkit V. Survey of satisfaction and complaint of customers of laboratory service, King Chulalongkorn Memorial Hospital. *Songkhanagarind. Med. J.* 2002 ; 20: 85-89.

[2]   Quality control in the laboratory. *Med. Klin.* 1974 Aug 2;69(31):1287-96.

[3]   Cooper GR. Quality performance in clinical pathology. Prog Clin Pathol. 1970;3:1-71.

[4]   Prier JE, Sideman L, Yankevitch IJ. Clinical laboratory proficiency testing. *Health Lab. Sci.* 1968 Jan;5(1):12-8.

[5]   Burke DS. Review of laboratory proficiency. *Infect. Control Hosp. Epidemiol.* 1988 Aug;9(8):365.

[6]   Abu-Amero KK. Overview of the laboratory accreditation programme of the College of American Pathologists. *East Mediterr. Health J.* 2002 Jul-Sep;8(4-5):654-63.

[7]   Aoyagi T. ISO 15189 medical laboratory accreditation. *Rinsho Byori.* 2004 Oct;52(10):860-5.

[8]   Kubono K. Quality management system in the medical laboratory--ISO15189 and laboratory accreditation. *Rinsho Byori.* 2004 Mar;52(3):274-8.

[9]   Bachner P, Howanitz PJ. Using Q-Probes to improve the quality of laboratory medicine: a quality improvement program of the College of American Pathologists. *Qual. Assur. Health Care.* 1991;3(3):167-77.

[10] Clinical pathology consultation improves coagulation factor utilization in hospitalized adults. *Am. J. Clin. Pathol.* 2003 Dec;120(6):938-43.

[11] Kondoh T, Kanno T. The assessment of quick turnaround service in clinical laboratories. *Rinsho Byori.* 1995 Oct;43(10):1038-43.

[12] Kumasaka K. Postgraduates' training as laboratory physicians/clinical pathologists in Japan--board certification of JSLM as a mandatory requirement for chairpersons of laboratory medicine. *Rinsho Byori.* 2002 Apr;50(4):353-7.

[13] Yanai M. Analysis of on-call consultations with clinical pathologists--identification of customer's satisfaction. *Rinsho Byori.* 2000 Sep;48(9):837-42.

Chapter II

Minimizing Errors in Laboratory Medicine and Effect on Patient Care

Abstract

Accuracy and precision of the laboratory result is the main aim of all laboratory analyses. Although errors are totally unwanted events in medical laboratory they usually occur. The errors in laboratory medicine can be either sporadic or systematic errors. Also, the errors can occur in any phase of laboratory cycle, preanalytical, analytical or post analytical phase. How to minimize the errors in laboratory medicine is the present focus of all medical laboratories. Several quality systems are proposed and can be applied for minimizing errors. It should be said the the highest attempt should be use to reach the level of zero error in laboratory medicine. A great usefulness of the highest attempt to conquer the error is equal to the usefuless of the patients. A significant reduction of misdiagnosis and mistreatment due to the errors of laboratory results can be expected.

Keywords: error, laboratory, effect.

Introduction

Wiwanitkit proposed in the published paper in BMC Clin Path that "Quality is the heart in management of all laboratories. Due to the laboratory quality cycle, reliability cannot be achieved in a clinical laboratory through the control of accuracy in the analytical phase of testing process alone. There should be a certification on the whole laboratory, but not on single analytical process." This can reflect the importance of quality in laboratory medicine. (Details are already mentioned in the first chapter of this book. According to the Rumsfeld's speech, "As we known there are known knows. There are things we know we know. We also know there are also known unknowns. That is to say we know there are some things we do not know. But there are also unknown unknowns, the ones we don't know we don't know." This speech can be the good idea for the beginner to realize the error in laboratory medicine. It can be seen that there are three facts: a) things that we know, b) things that we know we do not know and c) things that we have never know that we don't know. The last fact will bring error because we have never realized on it. The human beings can avoid the error but we have to manage it. Concerning the classical Chinese proverb "If you know, say you know and if you don't know, say you don't know. However, it still be the problems of the things that we have never know we don't know.

Every laboratory analysis has a possible degree of error associated with it. The error might derive from the analytical device and from the skill of the person doing the laboratory analysis. Errors can be seen in any phases of laboratory cycle: pre-analyical, analytical and post analytical phase. This means that we face up with risk for error at any time of laboratory analysis. Before jumping into the next step, we have to know the details of laboratory quality cycle. As already noted in the first chapter, the laboratory cycle has three phase, preanalytical, analytical and postanalytical phases. The preanalytical phases include the processes of patient preparation and specimen collection, transportation, sample receipt and assessment as well as preparation. The analytical phase includes actual analysis. The post analytical phase covers reporting and record keeping.

Natural History of Errors in Laboratory Medicine

1. Epidemiology

There are many reports on errors in medical laboratory. For example, Wiwanitkit (2001) [1] reported that the total errors in the laboratory was equal to 0.11 %, Plebani and Carraro (1997) [2] reported the sum of errors in their laboratory equal to 0.47 % and Carraro and Plebani (2007) further reported that the overall rate of error in their laboratory was equal to 0.31 % [3]. It should be noted that "Pre-analytical error is the most common kind of error in laboratory analysis [1]" Wiwanitkit (2001) [1] reported the rate of pre-analytical error in his laboratory equal to 84.5 % of all errors, Plebani and Carraro (1997) [2] reported the rate of pre-analytical error in their laboratory equal to 68.2 % of all and Carraro and Plebani (2007) [3] further reported the sum of 61.9 % pre-analytical error in their laboratory when compared to overall laboratory errors. However, Kazmierczak and Catrou (1993) [4] reported that preanalytical (0 %)and postanalytical (0.46 %) error were not major sources of laboratory error. This discordant report is very interesting. After careful consideration, it can be seen that this report has a pitfall in study design and collection of data.

2. Fade of Error

It can be seen that there are some reports on the fade of error. Plebani and Carraro [2] mentioned that "Most of the laboratory errors did not affect patients' outcome. However, some laboratory errors were associated with further inappropriate investigations, thus resulting in an unjustifiable increase in costs and those error could be associated with inappropriate care or inappropriate modification of therapy in the worst cases." Luckily, only a few percentage of error generates problem however problem is still problem and need concern. Bonini mentioned that [5] "The large heterogeneity of literature on laboratory errors together with the prevalence of evidence that most errors occur in the preanalytical phase suggest the implementation of a more rigorous methodology for error detection and classification and the adoption of proper technologies for error reduction. " This report also confirms the necessity to mange the error in laboratory medicine.

Preanalytical Error

There are several kinds of preanalytical errors. Wiwanitkit [1] reported that "Of all preanalytical errors, 95.2 % originated in the care units. All preanalytical errors, except for 1.15 % relating to the laboratory barcode reading machine, were due to human error. " The error can be detectable from all types of medical wards with the same high prevalence rate [1]. Therefore, it is needed to have quality control on these items: a) specimen collection site, b) venipuncture clinic setting, c) plebology training, d) standard practice protocol for specimen collection, e) quality system and f) ethics of practitioner. Rate of preanalytical errors is a quality indicator of medical laboratory. Therefore, a systematic record is necessary. A standard protocol should be set. So how to practice? The good protocol must be set. To help the reader, the author will present the experience which is already published in BMC Clin Path [1] "Briefly, all medical technologists in all units of the laboratory were asked to pay maximal critical attention to all received requests. These personnel were provided with a special notebook in which any "suspect" sample was recorded, together with all pertinent information. Then consultation to the head of medical technologist of the unit was done. The head of medical technologist rechecked and reviewed all reported cases before making final decision. In cases that the preanalytical mistakes were made from final decisions, they were recorded into the specific record form." It can be seen that specific record form must be designed and set. In the record form, there must be the parts for date, time, sources of request (for giving feedback), types and details of identified mistakes, primary corrective action (necessary for following up, setting of further preventive action and setting of secondary correction if needed) and signature of the recorder. Considering the types and details of identified mistakes, it can be physician's order miss, patient misidentification, specimen collected in incorrect quantity, inappropriate container used, inappropriate specimen's quality, specimen lost during transportation and etc. According to the study of Wiwanitkit, the inappropriate quantity of specimen is the most common kind of preanalytical mistake [6 - 8]. Indeed, the author has ever examined the newly graduated physician on the topic of specimen collection and found that most of the new physician cannot tell the proper quantity of specimen to be collected for many specific kinds of laboratory investigations [9].

Pre-Preanalytical Error

It should be noted that there might be another kind of error, pre-pre-analytical error. Any laboratory analysis in medicine has to be requested from the physician Following the concept that we have already discussed in the earlier part of the article on know and do not know, physician is a human therefore he/she can produce error. So what is the error for this case ? Loposta and Dighe [10] said that "A survey of physicians who use the clinical laboratory demonstrated that the largest number of test ordering errors appear to involve physicians simply ordering the wrong test." Errors can be due to request without indication, error in writing laboratory request form, error in patient identification and etc. It should be noted that incomplete laboratory request form writing is the major error found in laboratory requests. Medical personnel should be more careful in writing request forms and in specimen collection. It is noted that 63 per cent of the request forms were incomplete due to omissions, mistakes and use of non-standard abbreviations. Many errors were observed in aspects of the time that specimens were collected, diagnosis and patient identification.

Analytical Error

Analytical errors are the classical focus of all laboratories for a long time. Internal quality control and external quality assessment are the two main standard practice to reduce analytical errors [1]. However, the main problem is "Is this the best way?" Quality control is the core concept for analytical phase. Quality control includes operational techniques and activities used to fulfill requirements for quality. By definition, internal quality control (IQC) is set of procedures for continuously assessing laboratory work and the emergent results; immediate effect, should actually control release of results. This has to be done for all investigated test within the medical laboratory (as mentioned in the first chapter of this book). Quality assurance is another activity to be done. Quality assurance (QA) is planned and systematic activities to provide adequate confidence that requirements for quality will be reached and QA should and must include IQC, EQA, pre-analytic phase, test standardization, post-analytic phase, management, and organization. The whole process need to be assured.

There are four terms to be mentioned before studying of error in analytical phase.

- Accuracy means the agreement between a measurement and the true or correct value.
- Precision means the repeatability of measurement. It does not require us to know the correct or true value.
- Error means to the disagreement between a measurement and the true or accepted value. There are two kinds of error. The first, so called type 1 error, is statistical fluctuation in the measured data due to the problem of precision of the analytical device. This is also known as random error. The second, so called type 2 error, is inaccuracies when reproducible that tends in one direction. This can be named as systematic error.
- Uncertainty means an interval around that value such that any repetition of the measurement will produce a new result that lies within this interval. It should be noted that uncertainty is not equal to error but can contribute to error. In addition to error, every laboratory analysis has a possible degree of uncertainty associated with it. The uncertainty might derive from the analytical device and from the skill of the person doing the laboratory analysis. For sure, uncertainty can be the source of error.

To determine the uncertainty of the test becomes the needed practice. There are many references for determination of uncertainty (such as that of ANSI/NCLS, UKAS and etc.) that can be selected. Source of uncertainty in analytical phase can be due to many things. Weighting procedure is the common source. The weighting can be problematic and cause the problem. The other sources include purity of reagent and control, volume (pipette calibration is recommended), temperature (thermometer calibration is recommended) as well as analytical instrument (calibration is indicated). At present, the estimation of uncertainty efforts have been a laboratory issue and have helped to improve laboratory performance. It is also a rule of many quality system including ISO 17025 for laboratory.

Post-Preanalytical Error

Post analytical error takes a minor part of overall laboratory error in medical laboratory. However, this is still the problem to be managed. It should be noted that routine quality control and quality assurance process in analytical phase do not prevent post-analytical error. Human error is common. Incorrect validation is the best example. Post analytical error also includes misinterpretation of results. Laposata and Dighe [10] wrote that "The provision of an expert-driven interpretation by laboratory professionals resulted in improvements both in the time to and the accuracy of diagnosis. A survey of the physician staff has shown that in the absence of such an interpretation, for patients being assessed for a coagulation disorder, approximately 75% of the cases would have involved some level of test result misinterpretation." This can imply the problem of interpretation in the post analytical phase.

How to Manage the Problem of Error

How to manage the problem of error can be easily answered after the knowledge on the already mentioned topic. It can be easily said that management must include a) ensure quality of overall process, b) systematic detect and reduce errors, c) standard practice guideline, d), multidisciplinary approach, collaboration (ward – laboratory), d) surveillance and monitoring and e) continuous quality improvement (CQI). There are many quality management systems. Here, the author will briefly mentions for the two commonly used systems. First, hospital accreditation (HA) focuses on specific medical aspect. Clinical guideline, patient care team, ethics, practice guideline and etc. The other is ISO system (9002, 14000, 17025, 15189) which is adapted from industrial activity. ISO15169 is the specific version for medical laboratory.

The essentials of any quality management systems include a) process control, b) validation, quality control, proficiency testing, specimen management and etc., c) information management, d) control of documents and records, e) incident or occurrence management and f) preventive and corrective action.

Finally, the author would like to mention for "total automation" which is a new tool to reduce the error in laboratory. It brings reduction of human error. Pre-analytical module is the best example.

Effect of Laboratory Error Control on Patient Care

As previously noted, some errors brought to incorrect treatment, therefore, the best advantage of laboratory error control is to reduce those episodes. Reduced error means reduced necessity to repeat analysis and this has economical impact. Reduced error reduces risk of both clinician and technology for medical sue. The low incidence of error helps clinician rely on the laboratory report that he/she gets. Zero error is the benchmark. This must be possible. Please do not think that error is not serious. Can we trust on the plain that might pose error for traveling? Everyone must say no therefore this is the reason for setting the system to control error in medial laboratory. It should be noted that quality costs but no quality cost more especially in cases that the patients die because of laboratory error.

As evidence, Memeyer and Winkelman wrote in JAMA [11] that "In physician office laboratories where prothrombin time test volume is below 40 per month, the odds that a tested patient will experience a stroke or an acute myocardial infarction are up to 1.96 and 3.43 times greater, respectively, than for a similar patient tested in a commercial laboratory." Memeyer and Winkelman concluded that examining patient outcomes subsequent to clinical laboratory testing could be a useful tool for clinical laboratory quality assurance [11]. However, the most important factor is not automation or system but human factor. How to create "Quality culture in mind" is the focus.

References

[1] Wiwanitkit V. Types and frequency of preanalytical mistakes in the first Thai ISO 9002:1994 certified clinical laboratory, a 6 - month monitoring. *BMC Clin. Pathol.* 2001;1(1):5.

[2] Plebani M, Carraro P. Mistakes in a stat laboratory: types and frequency. *Clin. Chem.* 1997 Aug;43(8 Pt 1):1348-51.

[3] Carraro P, Plebani M. Errors in a stat laboratory: types and frequencies 10 years later. *Clin. Chem.* 2007 Jul;53(7):1338-42.

[4] Kazmierczak SC, Catrou PG. Laboratory error undetectable by customary quality control/quality assurance monitors. *Arch. Pathol. Lab. Med.* 1993 Jul;117(7):714-8.

[5] Bonini P, Plebani M, Ceriotti F, Rubboli F. Errors in laboratory medicine. *Clin. Chem.* 2002 May;48(5):691-8.

[6] Wiwanitkit V. Errors in laboratory requests in the In-Patient Department, King Chulalongkorn Memorial Hospital. *Chula Med. J.* 1998; 42(9): 685-693.

[7] Wiwanitkit V. Errors in laboratory requests in Chulalongkorn Hospital. *Maharat Nakhon Ratchasima Hosp. Med. Bull.* 2000; 24(2): 83-90.

[8]   Wiwanitkit V. Surveillance of non-conforming requests and specimens in King Chulalongkorn Memorial Hospital according to ISO9002 program, Medicine Beyond Frontiers the 42 annual Scientific Congress (Thai Physicians Association of America Faculty of Medicine Chulalongkorn University and King Chulalongkorn Memorial Hospital June 25-29 2001) (Abstract). *Chula Med. J.* 2001; 45(6): 534.

[9]   Wiwanitkit V. A knowledge survey of medical students about rational tube preparation. *Chula Med. J.* 2001; 44(5): 349-354.

[10]   Laposata M, Dighe A. "Pre-pre" and "post-post" analytical error: high-incidence patient safety hazards involving the clinical laboratory. *Clin. Chem. Lab. Med.* 2007;45(6):712-9.

[11]   Mennemeyer ST, Winkelman JW. Searching for inaccuracy in clinical laboratory testing using Medicare data. Evidence for prothrombin time. *JAMA.* 1993 Feb 24;269(8):1030-3.

Chapter III

Developing Patient Safety Culture in Clinical Laboratory

Abstract

There are several quality systems for medical laboratory at present. One of the high level system is the system focusing at the safety. The safety is basic requirement in all medical laboratories. Safety protocols should be set for all processes in medical laboratory. However, it should be noted that the safety should be extended to the patients. How to developing patient safety culture in clinical laboratory is an important point to be discussed. The total quality management in all steps, preanalytical, analytical and post analytical phases, must be used. In addition, the connection steps between laboratory and users (patients and medical wards) must also be managed. It should be kept in mind for all practitioners in clinical laboratory that they should prevent any accidents or errors for the patient as for themselves.

Keywords: patient, safety.

Introduction

The laboratory diagnosis is an important procedure in any clinical unit. It helps physician reach the solution for the queried disorders of their patients. However, poor quality of laboratory diagnosis cannot bring succeed but can bring the problem for users. Laboratory safety is a basic concept that every technologist must know and correctly practice. This starts from the safety of the practitioners. Basic laboratory safety rule include a) laboratory Hygiene, b) sharps safety, c) safety equipment, d) fire safety and e) chemical safety. However, it must generalize to the patients.

Safety for the Patients

Important laboratory users are patients. The patients must receive the most safe service. The concepts must be applied to all phase of laboratory analysis.

A. Preanalytical Safety

1. . Rational laboratory request (physician's role)

This means request with indication, without contraindication, informed consent and followed the concept of "First do no harm".

2. Safe patient preparation (Physician, nurse and technologist's role)

This means used of standard practice guideline. Patient identification must be done with special carefulness on infantile, pediatric and elderly groups who are dependant to care takers.

Clear communication with the patient must be done. Re-check for preparation with the patient before specimen collection is necessary.

3. Specimen collection (Physician, nurse and technologist's role)

This means having good practice guideline. All equipment passes sterility which is according to the rule of hygiene. Infection control is also recommended for some special cases such as immunocompromised hosts who are on chemotherapy. Universal precautions can be used.

4. Transportation (health care workers and technologist's role)

This also means having good practice guideline and control for good transportation.

Reconfirm of specimen for patient's identification and quality before further analysis is suggested.

B. Analytical Safety

Good analytical practice, good quality control and assurance are required. Although the error in analytical phase is not common it usually has a high impact on the patient management.

C. Postanalytical Safety

1. Rational laboratory interpretation (physician's role)

Basic knowledge requirement is needed. Re-identification of the patient is suggested.

Careful focus on unexpected aberrant laboratory result is recommended.

2. Proper result reporting system

This must include the topic on patient secret and patient's right. Good result reporting system is needed.

D. Information Management Safety

Wang and Ho wrote [1] that "A direct interface of the instruments to the laboratory information system showed that it had favorable effects on reducing laboratory errors."

Case Scenarios

1. Prenalytical Phase

1.1. Specimen Collection

Case 1. Sudden death in venipuncture clinic due to hypoglycemia [2]

This case is a case of an old diabetic patient receiving venipuncture at the venipuncture clinic. When performing venipuncture, the nurse in charge observed that the patients was unconsciousness. The nurse tried to examine the pulse and no pulse could be detected. The nurse called for help and the patient was brought to the emergency room and ended up with death. The rooted cause analysis showed that prolonged fasting in elder with diabetes mellitus and no specific protocol for management of this specific group of patient are the causes. This case implies importance of first AID training and CPR tool.

Case 2. Fainting and falling after venipuncture [3]

This a case of an old male visiting to the health unit to get the venipuncture. When performing the venipuncture in a small room, the patient suddenly felt faint and fell down on the floor. The patient had his head injury that required surgical management.

The rooted cause analysis showed that poor architecture setting, lack of protocol for management, no first AID training are all things bringing to this episode. This case implies importance of good venipuncture chair design, first AID training, bed and necessary primary drugs for the patients [4].

Case 3. Hematoma and infection at the venipuncture site

A patient sent the complaint to the laboratory manager that he had got the brusing skin lesion at his elbow at the venipuncture site that he got last week. He felt a lot of pain and noted that the area of bruising becomes redness and hot. The rooted cause analysis showed that poor venipuncture practice and poor hygeine in the venipuncture clinic were identified. Infection control is needed for all settings.

Case 4. Physician increased dosage of antidiabetic drug and bring unwanted side effect to the patient

This is a case of an old diabetic patient presenting to the physician with the problem of bleeding episodes. He had epistaxis, subconjunctival hemorrhage and bleeding per gum. He noted that he just got this signs and symptoms after he got the increased dosage of his antidiabetic drug from previous dosage 500 mg per day to 1850 mg per day. The tracing back shows that the phlebotomist in charge did not confirm for actual fasting (no water but pepsi !) of the patient and the physician in charge also did not re-confirm. This implies both errors in both preanalytical (specimen collection) and postanalytical (interpretation) phases. Protocol to manage unexpected laboratory results, special care for patients with aberrant laboratory results are required [5].

1.2. Transportation

Case 1. Accidental mislabeling during transportation

Accidental mislabeling during transportation can be seen. Label of a patient exchange with the other during transportation process by health care worker from ward to laboratory can be the cause of the episode. Double checking system, record form, re-identification with request form is required.

1. Analytical phase

Case 1. A cluster of proteinuria in screening for general university student

A cluster of proteinuria in screening for general university student is noted.

Luckily, the physician in charge observe this episode and notify back to the laboratory. Rooted cause analysis show out-of-date urine test strip. However, the laboratory has to pay for the extra-tests.

Case 2. A case of reported thick and thin film malaria negative with malaria

A case of reported thick and thin film malaria negative with malaria was reported in the incident report. Luckily, the physician in charge finally made diagnosis of malaria by performing blood smear examination by himself, however, this is delayed diagnosis. The non-experienced technologist did not detect the malarial inclusion Training of technician, certification and limitation of practice of technologist are necessary.

2. Postanalytical phase

Case 1. Accidental exchanging of blood gas results during reporting

Blood gas result of a patient exchange with the other during carrying reports by health care worker from laboratory to ward was observed. Luckily the physician in charge observed this aberrant result. Double checking system, record form, re-identification with request form, usage of information technology to help reporting are all suggested [5].

Conclusion

In conclusion, all contribute to patient safety culture. The aim of any medical service is to relief the unhappiness of the patients. It is no good that the patient dies despite complicated medical process. Think that patient is your cousin and you will be and can be a patient in one day in the future. Safety is not a difficult issue. It is a basic issue based on common sense.

Quality goes along with safety. No quality usually brings errors and accidents. Incidence report system must be set. "A laboratory incident report classification system can guide reduction of actual and potential adverse events.[6]" Aoston et al [6] also reported that "73 % of incidents were preventable (score, 3 or more). Of overall preventable incidents, 30 % involved cognitive errors, defined as incorrect choices caused by insufficient knowledge, and 73% involved noncognitive errors, defined as inadvertent or unconscious lapses in expected automatic behavior. " This refers to the fact that incident can be preventable in some scenarios and this should be done. In addition to incident report, satisfactory and complaint monitoring can help identify the problem of service. Corrective and preventive action are altogether also useful.

References

[1]   Wang S, Ho V. Corrections of clinical chemistry test results in a laboratory information system. *Arch. Pathol. Lab. Med.* 2004 Aug;128(8):890-2.

[2]   Wiwanitkit V. Case of sudden death in venipuncture clinic. 2004; 19 (4): 193.

[3]   Wiwanitkit V. A case study on patient management after venipuncture. *Chula Med. J.* 2006 ; 50(12): 869-872.

[4]   Wiwanitkit V. Modern concepts for venipuncture clinic setting. *Chula Med. J.* 2001; 45(5): 465-471.

[5]   Wiwanitkit V. Abnormal laboratory results as presentation in screening test. *Chula Med. J.* 1998; 42(12) : 1059-1067.

[6]   Astion ML, Shojania KG, Hamill TR, Kim S, Ng VL. Classifying laboratory incident reports to identify problems that jeopardize patient safety. *Am. J. Clin. Pathol.* 2003 Jul;120(1):18-26.

Chapter IV

Quality Management: Example by ISO System

Abstract

ISO is the wildely used new quality system in the laboratory. Medical ISO can be widely applied in all kinds of medical laboratory. It can also be applied to immunology study. Based on this system, the harmonization and standardization of immunology laboratories can be expected. Some discussion on the usefulness of the medical ISO and its application in laboratory cycle of immunology laboratory are presented in this article. This can be the good example of quality management by ISO system.

Keywords: ISO, immunology, laboratory.

Introduction

Improving from the basic concepts of quality control, quality assurance and total quality management, Internation Organization for Standardization (ISO 9000) is used world-wide in the presented day in many medical laboratories. ISO system has been developed on the basis for establishing quality system in service. Because medicine is a type of service, therefore, it should follow this system. Any aspects in medicine need controlling and improvement by this system. There are several present applications of ISO system in laboratory medicine, such as systems in administration, analysis and medical instrumentation production. So this system should be understoond by all medical personals in order to control and improve their medical service for the good laboratory and good clinical practice. In order to jump into the international level, the international standard must be implemented and maintained. The international standard basically provides quality management to cover all laboratory processes from pre-analytical, analytical, and post-analytical processes. Owing to the author's experience, it is found that not only the standard could be applied to assure the laboratory quality system but it also could be used to support us create a new quality culture of continuous quality improvement. In realization of the usefulness of the standard, the new ISO

15189: 2003 edited from combination of ISO/IEC 17025 and ISO 9001: 2000 has been selected as the present best ISO system for medical laboratory. To help the reader better understand on ISO, the author hereby presents this article on example of quality management by medical ISO. The model of ISO15189 for medical laboratory will be hereby proposed and hope to be the basic idea for readers in implementation of the appropraite system for their specific laboratories.

Application of ISO for Medical Laboratory

In laboratory medicine, the two main important aims are good service and good quality. In the previous day, good quality usually focused mainly on good quality control and assessment in the analytical process. However, reliability cannot be achieved in a clinical laboratory through the control of accuracy in the analytical phase of the testing process alone but it requires additional concerns on several topics. Due to the good governance concepts, accountability of the whole laboratory process is the main focus of present general consideration in laboratory medicine. There must be a certification on the whole laboratory, but not on single analytical process. Precision and accuracy of analyses are not only detected by the analytical procedure but also by pre-analytical factors as well as post-analytical factors. In addition, presently, in case of a medical laboratory "good medical laboratory services" is preferred worldwide. Total quality management (TQM) of laboratory services is a concept globally used at present [1]. Since, TQM concentrates not only on analytic performance and organizational issues, including specimen collection, reporting, and interpretation of results, but covers also on the benefits to society related to the use of specific laboratory investigations in prevention, early detection, and therapy monitoring, as well as on outcome measures, the good service and quality of the laboratory can be imagined. Similar to other medical laboratories, immunology laboratory needs the TQM. A prerequisite to TQM are international and national standardization programs for the establishment of optimized and standardized methods, as well as for the development and evaluation of appropriate reference materials. ISO series have been widely used for clinical laboratory for a few recent years. The main considerations for ISO 15189 series cover service as well as technical processes [3]. The considerations for the service processes are not different from the other ISO series, covering the policies, resource administration, corrective actions and preventive actions. While the considerations for the technical processes are specifically according to general medical laboratory quality principles. The three phases of the medical laboratory cycle are also included.

Upon completion of application of this ISO series to the medical laboratory, participants must learn the concept and practice of the standard requirements, and have to adjust the standards to fit the standard to their workplaces and to plan specific implementation to help improve the quality of their laboratory testing. The participants should focus on not only the laboratory process but also the patient outcome and focuses on the total service of a medical laboratory. Similar to the other ISO series, the settlement of the ISO 15189 in a medical laboratory requires the cooperation from all medical personnel who practice in the laboratory. The head or director of the laboratory should have leading behavior, knowledge and good resource management skill. The preparation for the document is another important phase since the auditing visit usually bases on the document as evidences. The first concerned point is about the administration level of the laboratory. Usually, the administrator of the laboratory does not well prepare for the future planning for resource management. Routinely, proper resource management should cover all 4'M: man, material, money and management.

The other aspects in services processes that oftenly recieve non-conformation to the standard are usually related to the statistical collection of the laboratory. The laboratory usually has no supportive evidences for corrective and preventive actions. Considering the corrective actions, the surveillance system is necessary. Monitoring for basic important key performance index including satisfaction, complaint, incident report, non-conforming specimen, non-conforming laboratory result, instrument error, personal absence and turnaround time is recommended for all medical laboratories. Using this monitoring system, the laboratory can effectively the corrective action to the detected problem.

ISO for Laboratory Cycle in Immunology

Basically, ISO15189 is globally used for clinical chemistry laboratory. However, it can also be set for the clinical immunology laboratory setting. As a rule, the pre-analytical phase error in immunology study shows higher incidence than the other left two phases. The good quality management for this specific phase is necessary. Hence, the control of specimen is recommended. In case that the specimens are not collected and controlled by the laboratory, the specific protocol should be distributed to the collection sites. Indeed, although the specimens are collected by the laboratory, such specific specimen controlling protocol must be developed. The protocol should described the necessary details of the test, the collection technique, patient preparation technique, transportation, technique, precaution, turnaround time, reference, price and other necessary contact detail. It is routinely mentioned that wrong collecting and sending of sample are main sources of errors for immunology laboratory [3]. In addition, wrong specimen collection can bring incorrect report and it can be a big problem such as the misidentification of HIV infected case [4]. The main point that the ISO15189 for clinical immunology is different from that for clinical chemistry is the management of some specific immunological tests. Consultation on evaluation of the immune system have to be available to help the physician taking care of patients with or suspected with primary immunodeficiency. When the specimen reaches the laboratory, before analysis, another important basic process is specimen preparation. In this phase, the laboratory must look for the possible non-conforming specimens, especially for clot and hemolysis within the recieved specimens. If the non-conforming specimen can be observed, rejections accompanied with suggestion for corrective method should be provided to the client.

As a rule of ISO 15189, all analytical methods have to be documented and traceable. All material including reagent and equipment should be good controlled and monitored. All personnel must be certified for their specific works or jobs: ones who relate to the result validation should be registered and authorized. Concerning the analytical methods, the common pitfalls usually relates to lack for method evaluation before and after start of laboratory service. The lack of reference values for the laboratory is another common pitfall that can be detected in the medical laboratory service. Concerning the reagent, the evaluation before and after usage of each lot of reagent is needed. The results of evaluation have to be recorded and kept as evidence for traceable. The labeling of in use regent is necessary; including opening data, expired date as well as lot number. Concerning the equipment, all equipments have to have their own specific history, indicating manufacturer, supplier, history of fixing, maintenance, setting and movement. All equipments needed to be certified and calibrated: it should be noted that calibration has to be performed by the professional organization, which can be traceable to the international standard. In addition, all analytical processes have to be under the internal quality control and external quality assessment program [5], which international proficiency test is preferred. However, in case that the resource is limited, inter-laboratory assessment is allowable and acceptable. All documented relating to the quality control and assessment program should be collected and kept as evidence for traceable. After complete analysis, the laboratory has to be systematic validated by the registered authorized medical personnel. The personnel who practice in validation must sign for their responsibility. Repeated recheck for every step is recommended. In case that the laboratory information system (LIS) is available for use, the validation and maintenance for the LIS as well as recheck for the agreement of the results in intra-laboratory and extra-laboratory screens are necessary. Another point that the ISO15189 for clinical immunology is different from that for clinical chemistry is the management of laboratory result confidentiality. Specific procedures for management of secret results such as Anti HIV results must be addressed.

References

[1]   Kuwa K. Laboratory accreditation and proficiency testing. *Rinsho Byori.* 2003;51:449-55.

[2]   Burnett D. ISO 15189:2003--quality management, evaluation and continual improvement. *Clin. Chem. Lab. Med.* 2006;44:733-9

[3]   Gershgori EG. Wrong collecting and sending of saliva samples as a source of expert error. *Sud. Med. Ekspert.* 1966;9:16-7.

[4]   Krombach J, Kampe S, Gathof BS, Diefenbach C, Kasper SM.. Human error: the persisting risk of blood transfusion: a report of five cases. *Anesth. Analg.* 2002 ;94:154-6

[5]   Lock RJ. My approach to internal quality control in a clinical immunology laboratory. *J. Clin. Pathol.* 2006;59:681-4.

Chapter V

Cost Study in Laboratory Medicine: Example

Abstract

Cost is a big concern in present world economics crisis. Cost study in laboratory medicine is important. This article will present examples of cost studies in laboratory to help the readers better understand this topic.

Keywords: cost, laboratory.

Example 1: Comparative Study of the Cost-Effectiveness between Macroscopic Slide Cell Agglutination and Microscopic Agglutination Test for Diagnosis of Leptospirosis

1. Introduction

Leptospirosis is a bacterial disease of the genus *Leptospira* that affects humans and animals. It is caused by bacteria. In humans it causes a wide range of symptoms, and some infected persons may have no symptoms at all. Symptoms of leptospirosis include high fever, severe headache, chills, muscle aches, and vomiting, and may include jaundice (yellow skin and eyes), red eyes, abdominal pain, diarrhea, or a rash. If the disease is not treated, the patient can develop kidney damage, meningitis (inflammation of the membrane around the brain and spinal cord), liver failure, and respiratory distress [1 – 2].

Outbreaks of leptospirosis are usually caused by exposure to water contaminated with the urine of infected animals [1 – 2]. Many different kinds of animals carry the bacterium; they may become sick but sometimes have no symptoms. Leptospira organisms have been found in cattle, pigs, horses, dogs, rodents, and many wild animals. Humans become infected through contact with water, food, or soil containing urine from an infected animal. In Thailand, the leptospirosis is still an important public health problem, with a recent outbreak in the Northeast in 1998 [3].

Many of these symptoms can be mistaken for other tropical diseases. Leptospirosis is confirmed by laboratory testing of a blood or urine sample. In Thailand, the diagnosis is mainly based on serological methods presently. The microscopic agglutination test (MAT) is a reference method for the detection of antibodies to leptospirosis. However, few laboratories can perform MAT because the maintenance of living organisms is required for this assay, and this creates a hassle of laboratory personnel to perform. Therefore, the alternative method as macroscopic slide cell agglutination (MAT) is introduced. However, there is no previous cost effectiveness evaluation comparing these two methods, hence, this study was performed

2. Materials and Methods

This study was designed as a cost effectiveness study. The main aim of this study was to compare the costs of macroscopic slide cell agglutination (MCA) and the microscopic agglutination test (MAT) for diagnosis of leptospirosis. Data on cost and effectiveness of the two methods were reviewed. We used King Chulalongkorn Memorial Hospital (KCMH), the Thai Red Cross Society Hospital as representative setting in Thailand. The KCMH is a tertiary hospital with more than 1,000,000 patients per year.

Cost in this study was defined as unit cost reported by the laboratory of the hospital and presented in baht (Table 1). The utility or effectiveness was derived from the prevalence of disease among the Thai population in a recent report, about 3 % [5]. A decision tree depicting the two methods strategies and associated probabilities, costs and utilities is presented in Figure 1. The probabilities for each path were derived from the detection characteristics derived from a previous validation study in this specific setting [4].

Table 1. Costs and utilities of MAT VS MCA for diagnosis of leptospirosis

Methods	result	Path probabilities*	Costs** (baht)	Utilities***
MAT	+	0.213	450	0.030
	-	0.787	450	0.970
MCA	+	0.066	350	0.030
	-	0.934	350	0.970

*The probabilities for each path were derived from the detection characteristics derived from the previous validation study in our setting [4]

** Cost in this study was set defined as unit cost reported by the laboratory of the hospital and presented in baht.

*** The utility or effectiveness was derived from the prevalence of disease among the Thai population in a recent report, about 6.5 % [4].

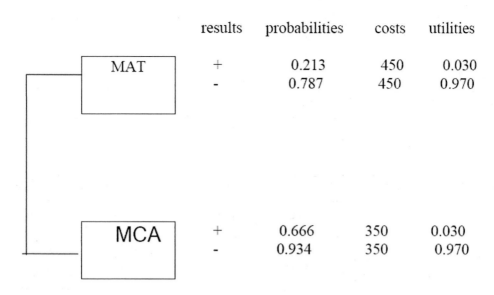

	results	probabilities	costs	utilities
MAT	+	0.213	450	0.030
	-	0.787	450	0.970
MCA	+	0.666	350	0.030
	-	0.934	350	0.970

Figure 1. Decision tree leptospirosis detection method strategies and associated probabilities, costs and utilities.

The results from each category were collected and analyzed using an economical statistical method. The expected cost of each strategy was derived by multiplying cost for branch with the probability of that branch and subsequently adding all such products derived from the branches of that strategy. Similarity the expected utility of each strategy was determined.

3. Results

The cost and utility of each screening method are shown in Table 1. The expected cost and utility,and cost – effectiveness, of each method are presented in Table 2. According to this cost effectiveness evaluation, MCA is more cost effectiveness of MAT.

Table 2. Cost-effectiveness analysis

Strategies	Expected cost (baht)	Expected utility	Cost-effectiveness* (baht)
MAT	450	0.770	584.42
MCA	350	0.908	385.46

* cost effectiveness = expected cost/ expected utility

** The expected cost of each strategy was derived by multiplying cost for branch with the probability of that branch and subsequently adding all such products derived from the branches of that strategy. Similarity the expected utility of each strategy was determined.

4. Discussion

Leptospirosis can be misdiagnosed because it may be confused with other common febrile diseases, which are common in Thailand as well [6]. The most accurate diagnosis of leptospirosis is based on the isolation of the causative organisms. However, leptospirosis culture takes up to 2 months and is truly useless for clinical practice. Therefore, the gold standard for diagnosis is based on the serological methods [7].

Of the various serological methods, two are frequently used in Thailand - MAT and MCA. These two serological methods have been evaluated for their diagnostic sensitivity and their sensitivities are similar [4]. Due to the present economical crisis in Thailand, however, not only the diagnostic sensitivity but also the cost effective of all tests are of concern.

According to this cost effectiveness evaluation, the overall cost-effectiveness of MCA is lower than the MAT (Table 2). It seems that the MAT method might be more cost effectiveness method in terms of actual cost per diagnosis, but if the turnaround time is also considered. MCA can provide a more cost-effective result with a faster turnaround time (the highest cost effective – turnaround time ratio) (Table 3). In conclusion, based on our setting in Thailand, the author recommends the MAT as the preferred method for determination of leptopsirosis in settings with a low workload, and MCA as the preferred method for determination of leptopsirosis in the settings with a high workload.

Table 3. Cost-effectiveness relative to the turnaround time

Strategies	Cost-effectiveness (baht)	Turnaround time (minute)*	Cost effectiveness - turnaround time ratio (baht/minute)
MAT	584.42	120	4.87
MCA	385.46	5	77.09

*Turnaround time was set as the overall time used for the analytical process based on our setting
**Cost effectiveness turnaround time ratio = cost effectiveness/turnaround time.

5. Conclusion

A cost effectiveness study was performed to compare the MAT and MCA for diagnosis of leptospirosis. It was found that MCA is more cost effectiveness than MAT, however, concerning the turnaround time as a second factor, MCA can provide a cost-effective result with the faster turnaround time (the highest cost effectiveness – turnaround time ratio).

Example 2: Cost Effective Analysis of Cultivation Method for Trichomoniasis

1. Basic Information on Trichnomoniasis

1.1. What Is the Importance of Trichomonas Vaginalis Infection?

Trichomoniasis is one of the most common treatable sexually transmitted disease worldwide [7]. It is associated with probably serious complications such as preterm birth and human immunodeficiency virus acquisition and transmission [7].

Trichomonas vaginalis is the causative agent of trichomoniasis. *T. vaginalis* was originally considered a commensal organism until the early of 1950s when the understanding of its role as a sexually transmitted infection (STI) began to be accepted [8]. *T vaginalis* is site specific for the genitourinary tract and has been isolated from virtually all genitourinary parts [8]. Asymptomatic disease is common in both men and women, hence, screening for disease is very important [8]. Various sociodemographic factors have been proved to relate to presence of *T. vaginalis*, and may be used to predict infection [8].

Among women, trichomoniasis may play a role in development of cervical neoplasia, postoperative infections, and adverse pregnancy results and as an important factor in atypical pelvic inflammatory disease and infertility [9]. Trichomoniasis has been related to vaginitis, cervicitis, urethritis, pelvic inflammatory disease (PID), and adverse birth outcomes [8]. Among men, trichomoniasis has accepted as a cause of nongonoccocal urethritis and as contributing to male factor infertility [9]. Since the infection continues to accumulate, it is important to find strategies to increase diagnostic efforts beyond traditional sexually transmitted disease clinic settings [9 - 10]. Schwebke said that despite being a readily diagnosed and treated STD, trichomoniasis was not a reportable infection and control of this sexually transmitted infection had received relatively little emphasis from public health STD control efforts [11]. Schwebke also noted that an appreciation of high rates of disease and of associations of trichomoniasis in women with adverse outcomes of pregnancy and increased risk for HIV infection brought a requirement for increased control efforts [11].

1.2. Diagnosis for Trichomonas Vaginalis Infection

Diagnosis of trichomoniasis is usually based on wet mount microscopy and direct visualisation, which are insensitive [8]. Culture for *T. vaginalis* is more reliable test but requires more complicated technique. Edwards said that a careful microscopic examination of vaginal secretions generally yielded the correct diagnosis, but atypical or recalcitrant disease deserved a confirmatory culture, as noninfectious inflammatory processes can produce similar symptoms [12]. DNA amplification techniques perform with good sensitivity, but are not already proved for diagnostic purposes [8]. In areas where diagnostic methods are limited, management of trichomoniasis is commonly as part of a clinical syndrome; vaginal discharge for women and urethral discharge for men [8].

Another simple test that can detect trichomonas infection is urinalysis.

In 1984, the consequence of omitting urine sediment microscopy in specimens with normal physicochemical testing was assessed in a retrospective review of laboratory and clinical data in 1,000 patients by Valenstein and Koepke [13]. In their work, the pathologic states of clinical significance most likely to be overlooked were Trichomonas infection and occasional cases of significant bacteriuria [13]. Berg et al also reported high prevalence of sexually transmitted diseases in women with urinary infections is USA [14]. However, Valenstein and Koepke noted that the medical benefit of performing urine microscopy in these two groups of patients was not clear and they recommended reserving microscopy for urine specimens that show physicochemical abnormalities [13].

In 2000, Patel et al performed a systematic review of literature indexed in MEDLINE of diagnostic studies that used Trichomonas culture as gold standard [15]. The sensitivity of the polymerase chain reaction technique (PCR) was 95%, and the specificity was 98% [15]. The sensitivity of the enzyme-linked immunosorbent assay in this report was 82%, and the specificity was 73% [15]. The sensitivity of the direct fluorescence antibody in this report was 85%, and the specificity was 99% [15]. Sensitivities of culture media were 95% for Diamond's, 96% for Hollander, and 95% for Cystein-Peptone-Liver infusion-Maltose (CPLM) [15].

1.3. Culture Media for Trichomonas Vaginalis

Cultivation of luminal protistan parasites has a very long history [16]. Several cultures media is developed, however, there are only two commercially available, ready-to-use culture media which are confirmed for clinical application by the Food and Drug Administration for clinical diagnosis of vaginal trichomoniasis: Kupferberg's STS and Diamond's medium (modified) [17]. Negm and el-Haleem found that the modified Diamond's media proved to be highly sensitive (80.95%) and the results of the culture were significantly associated with those of PCR [18]. The results of both wet mount and AO stain were not significantly associated [18]. The wet mount although commonly used, easy, rapid and inexpensive yet, it has low sensitivity hence negative cases should be repeated by either culture or PCR [18]. According to the study of Gelbart et al, Diamond's medium (Klaas modification), recommended by the Centers for Disease Control for the isolation of *T. vaginalis*, was compared in vitro to Kupferberg's (STS) medium [17]. In this study, growth studies using six fresh clinical isolates, all from different patients, showed that while generation time was about 6 hours in both STS and Diamond's, the period of exponential growth was appeared longer in Diamond's [17]. More important, in STS there was a 4-hour lag period during which the population significantly reduced prior to exponential growth and this did not occur in Diamond's medium [17]. Gelbart et al concluded that because Diamond's medium (modified) allowed more prolific growth over a shorter period of time, it was more suitable than Kupferberg's (STS) for detecting *T. vaginalis* in patients with vaginitis [17]. Gelbart et al also noted that Diamond's medium was found to be superior to both Kupferberg medium and wet mount [19]. Philip et al found that culture on modified Diamond's medium was more sensitive (98%) than the wet mount method (38%) in detecting *T. vaginalis* [20]. They proposed that the wet mount method was very insensitive for detecting *T. vaginalis* and was positive only in the patients yielding greater than 10^5 cfu/ml [20].

Considering Hollander media, this is an old media. It is limited used at present. Considering CPLM media, it is an alternative media that can be prepared in the medical laboratory at present. It is widely used in many developing countries. Recently, Saksirisampant et al introduced new modified CPLM media [21]. The new modified media is acceptable with lower cost comparing to standard CPLM media [21]. Abd el Ghaffar et al performed a comparative study between CPLM and TYM in order to compare their ability to isolate and to maintain the growth of *T. vaginalis* in the laboratory [22]. While both media were found to be similarly good in detecting the organisms in vaginal discharges, yet, *T. vaginalis* stocks were maintained for a longer time in TYM medium (one year), than in the CPLM medium (14 dyas) [2]. The yields of the parasites with different inocula subcultured and after different incubation periods were also detectd in the TYM medium [22]. Here the author will report a cost utility analysis of those tests in medical practice.

2. Materials and Methods

2.1. Alternative for Cultivation

As described, there are three common cultivation methods for confirmation diagnosis of trichomoniasis. CPLM, Modified CPLM and BBL

2.2. Cost Analysis

The cost in baht (1 US dollar = 41 baht) for performing each test was reviewed. The cost used was set as the price of each test at the reference laboratory in Thailand (Division of Parasitology, King Chulalongkorn Memorial Hospital, Bangkok Thailand).

2.3. Cost Utility Analysis

The cost for each alternative node for diagnosis of trichomoniasis is calculated. The utility of each method is defined as the mean increasing number of *T. vaginalis*, which is quoted on our recent study [4]. The cost utility analysis is then performed. The operative definition of cost utility is cost divided by the utility similar to other cost utility study.

3. Results

Cost of each alternative method for diagnosis of trichomominiasis are presented in Table 3. The cost per utility of CPLM is the highest and modified CPLM is the lowest.

Table 3. Cost and utility of each alternative node for diagnosis of trichomoniasis

CPLM		Modified CPLM		BBL	
Cost (baht)*	Utility (rate)**	Cost (baht)	Utility (rate)	Cost (baht)	Utility (rate)
8.11	10.47	2.13	54.33	2.97	11.23

** cost per test
* The utility of each method is defined as the mean increasing number of *T. vaginalis*, which is quoted on our recent study [21].

**Table 4. Cost utility analysis of each alternative
node for diagnosis of trichomoniasis**

Alternatives	Cost per utility
CPLM	0.77
Modified CPLM	0.04
BBL	0.26

4. Discussion

Trichomoniasis is perhaps the most common curable sexually transmitted disease worldwide. It is associated with potentially serious complications as already noted in the previous part of this article[11]. The immunology of a related organism, Tritrichomonas foetus, which causes disease in cattle, has been investigated to some extent, but more work is required for the human strain, *T. vaginalis* [11]. In addition, although trichomoniasis is easily cured with oral metronidazole, there is concern that the number of strains resistant to this antibiotic are increasing [11, 15]. It can be shown that the sensitivity of the cultivation is similar to PCR, therefore, it is accepted as a standard confirmation test for trichomoniasis. Here, the author performed an economical analysis for the three common cultivation tests widely used for diagnosis of trichomoniasis. Here, it can be shown that the cost per utility for modified CPLM is the least expensive choice. Therefore, this alternative is the best method for cultivation diagnosis for trichomoniasis, based on medical laboratory economics principles.

References

[1]   Vinetz JM. Leptospirosis. *Curr. Opin. Infect. Dis.* 2001; 14: 527-38

[2]   Levett PN. Leptospirosis. *Clin. Microbiol. Rev.* 2001; 14: 296-326

[3]   Tangkanakul W, Tharmaphornpil P, Plikaytis BD, Bragg S, Poonsuksombat D, Choomkasien P, Kingnate D, Ashford DA. Risk factors associated with leptospirosis in northeastern Thailand, 1998. *Am. J. Trop. Med. Hyg.* 2000;63: 204-8

[4]   Tirawatnapong S, Chirathaworn C. Macroscopic slide cell agglutination test for rapid diagnosis of leptospirosis. *Chula Med. J.* 1999; 43: 141 - 6

[5]   Sehgal SC. Global scenario of leptospirosis. Presented at the Joint International Tropical Medicine Meeting 2001, Bangkok Thailand. 8 – 10 August, 2001

[6]   Kobayashi Y. Clinical observation and treatment of leptospirosis. *J. Infect. Chemother.* 2001;7:59-68

[7]   Palmer MF. Laboratory diagnosis of leptospirosis. *Med. Lab. Sci.* 1988;45, 174 - 8

[8]   Schwebke JR, Burgess D. Trichomoniasis. *Clin. Microbiol. Rev.* 2004 ;17:794-803

[9]   Swygard H, Sena AC, Hobbs MM, Cohen MS. richomoniasis: clinical manifestations, diagnosis and management. *Sex Transm. Infect.* 2004 ;80:91-5.

[10]   Soper D. Trichomoniasis: under control or undercontrolled? *Am. J. Obstet. Gynecol.* 2004;190:281-90.

[11] Schwebke JR. Update of trichomoniasis. *Sex Transm. Infect.* 2002;78:378-9.

[12] Edwards L. The diagnosis and treatment of infectious vaginitis. *Dermatol. Ther.* 2004;17:102-10.

[13] Valenstein PN, Koepke JA. Unnecessary microscopy in routine urinalysis. *Am. J. Clin. Pathol.* 1984;82:444-8.

[14] Berg E, Benson DM, Haraszkiewicz P, Grieb J, McDonald J. High prevalence of sexually transmitted diseases in women with urinary infections. *Acad. Emerg. Med.* 1996;3:1030-4.

[15] Patel SR, Wiese W, Patel SC, Ohl C, Byrd JC, Estrada CA. Systematic review of diagnostic tests for vaginal trichomoniasis. Infect Dis Obstet Gynecol. 2000;8:248-57.

[16] Clark CG, Diamond LS. Methods for cultivation of luminal parasitic protists of clinical importance. *Clin. Microbiol. Rev.* 2002;15:329-41.

[17] Gelbart SM, Thomason JL, Osypowski PJ, Kellett AV, James JA, Broekhuizen FF. Growth of Trichomonas vaginalis in commercial culture media. *J. Clin. Microbiol.* 1990;28:962-4.

[18] Negm AY, el-Haleem DA. Detection of trichomoniasis in vaginal specimens by both conventional and modern molecular tools. *J. Egypt Soc. Parasitol.* 2004 ;34:589-600.

[19] Gelbart SM, Thomason JL, Osypowski PJ, James JA, Hamilton PR. Comparison of Diamond's medium modified and Kupferberg medium for detection of Trichomonas vaginalis. *J. Clin. Microbiol.* 1989;27:1095-6.

[20] Philip A, Carter-Scott P, Rogers C. An agar culture technique to quantitate Trichomonas vaginalis from women. *J. Infect. Dis.* 1987;155:304-8.

[21] 21.Saksirisampant W, Saksirisampant C. A using of modified Cystein-Peptone-Liver infusion-Maltose medium for cultivation of Trichomonas vaginalis. *Chula Med. J.* 2001; 45: 31-38

[22] Abd el Ghaffar FM, Azab ME, Salem SA, Habib KS, Maklad KM, Habib FS. Evaluation of two cultural media (CPLM & TYM) for isolation and maintenance of Trichomonas vaginalis stocks in the laboratory. *J. Egypt. Soc. Parasitol.* 1994;24:611-9.

Chapter VI

Problems of Phlebotomist

Abstract

Phlebotomy is the widely practiced medical procedure. This is the basic procedure for collection of venous blood for further laboratory analysis. Phlebotomist is the specific group of medical personnel who practice venipuncture. This article will focus on problem of phlebotomist.

Keywords: phlebotomist.

Tuberculosis among the Phlebotomist, an Experience from Thailand

Tuberculosis (TB) is a biological hazard for the medical personnel at present. The contact with the patients is the main cause of getting infection among the medical personnel.

Although completely eliminating the risk for transmission of M. tuberculosis in all health-care facilities may not be possible, adherence to the principles outlined in the CDC guidelines should reduce the risk to persons in such settings [1]. The phlebotomist is a type of medical personnel that has high risk to get TB from contact of the patients. According to the MMWR report, infection in phletomist is detected in the outbreak of TB in a community hospital in USA [2]. Here, the author performed a retrospective analysis on the infectious control report in a phlebotomy clinic of a Thai tertiary hospital.

The author reviewed the infectious control report of the phlebotomy clinic of King Chulalongkorn Memorial Hospital during year 2003. The incidence of TB among the phlebotomists is equal to 2 cases. The proportion of the affected phlebotomists to overall phlebotomist equal to 2: 18. Concerning the risk, the ratio of infection per phlebotomy services is equal to 2 cases: 72,000 blood collections. The two infected cases were prompt treatment by antituberculosis drug. Good outcome is derived. Specific protocol to control of respiratory infection was applied to this clinic. The glass block was used to separate between the patients and phlebotomists during blood collection. Follow up on year 2004, there was neither new case nor relapse of TB infection. The author hereby states that the infectious control and prevention for TB is necessary for every phlebotomy clinic. In addition, an annual check up program for the phlebotomist is also recommended.

A Case Study on Needle Stick Injury on a Phlebotomist

1. Introduction

Accident is a totally unwanted episode in medical laboratory [3]. There are several types of accident in laboratory. However, the most common type is accident during specimen collection [4]. Needle stick injury is the most problematic accidental injury in medicine [5]. This type of accident is considerred serious because it can be a cause of many consequent complications.

In this article, the anther reports a case study on needle stick injury on phlebotomist. A discussion on this case in several aspect in laboratory medicine was performed.

2. Case Study

A female phlebotomist performed a venipuncture on an HIV-infected patient laying on a bed. This patient was moved from the out - patient clinic to get the venipuncture at the phlebotomy clinic of the hospital. The patient on the bed was placed on a specific corner for venipuncture and curtain was used to shield this urea form third person. However, the patient's cousin can observe venipuncture practice.

During performing venipuncture, the phlebotomist was attacked backward by one of the patient's cousin. The accidental needle stick injury occurred. The phlebotomist got immediate first aid. She consulted to the infection control specialist and submitted an accident report. After obtaining antiretroviral drugs, this phlebotomist developed nausea - vomiting symptom. The physician in change allowed her to have a rest for 3 weeks. The followed up Anti HIV test results were all negative.

3. Discussion

Needle stick injury is common in general medicine. However, the incidence among phlebotomist, occupational worker who practice venipuncture every day, is rare. According to the author's laboratory record, there are about 100,000 venipuncture per year. This equals to about 10,000 venipunctures per phlebotomist per year. Of interest, according to strict laboratory standard, there had been no incidence of accidental needles stick injury for years. This incidence is an interesting problem.

This accident occurred due to third person. Although good protocol of venipuncture and patient preparation was set there was still a pitfall due to patient's cousin. Allowance of patient's cousin to observe clinical practice seems to be a controversial issue. If medical personnel allow cousin to observe specimen collection procedure, the cousin must be kept in appropriate position that cannot disturb the procedure.

In the case, the phlebotomist performed correct post exposure management. Reporting and consultation was performed [6 – 7]. This case was considered a serious injury, known seropositive source with piecing injury. The use of antiretroviral drug for secondary prevention is recommended. Indeed, a common main pitfall is no report or delayed getting antiretroviral drug. van Oosterhout et al reported that most of affected cases in their setting did not report the needle stick incidence [8]. Wig also highlighted the low awareness of postexposure prophylaxis measures amongst health care workers [9]. However, there was no problem at this stage in this case. An interesting problem in this one is adverse effect of antiretroviral drug administration. Kowalska et al reported that adherence to consultations and tests schedule was not good in health care workers [10]. Due to the severe side effect, the phlebotomist was advised to rest for many days. This affected the venipuncture unit to increased workload on other phlebotomists. This is in interesting forgotten complication of this accidental incidence in the laboratory.

Table 1. Points to be concerned in accidental injury in the phlebotomist

Steps	Concerns
1. before injury	• proper patient preparation, avoid movement range of the patient that can attack the phlebotomist to get needle stick injury • application of sharp or certain to protect venipuncture form others patients in case of laying down patient • Allowance of patient's cousin to observe in an appropriate site.
2. after injury	• Strict universals precautions practice • Use of safety protective instruments • Prompt wound care and first aid • Reporting system is need. No report can bring lost of right in case of seroconversion • Consultation to specialist of estimation for risk and getting antiretroviral drug if indicated. • Following up for side effect in case prescribed with antiretroviral drug • Root cause analysis and additional preventive action • Concern of social impact on affected case and effect to the causative patient (right of him/her to deny serological test, and other medical personnel (who have to take duty instead of affected medical personnel).

Preventive Medicine for Venipuncture

1. General Information on Occupational Exposure to Blood Borne Pathogens [11 - 12]

An exposure that can place health care at risk for many blood borne infections is defined as either of these scenarios

- a per-cutaneous injury (such as a needle-stick or cut with a sharp object) or
- contact of mucous membrane (such as exposed skin that is chapped, abraded, or afflicted with dermatitis)
- with blood, tissue, or other body fluids that are probably infectious .

Therefore, preventive medicine for blood collection procedure is very important. The opportunity statement for the occupational exposure to blood borne pathogens globally stated in laboratory medicine. It is well accepted that accidental injuries from needles used in healthcare and laboratory settings may result in transmission of bloodborne infectious pathogens to healthcare workers. The most likely etiologies of those accidental injuries include a) failure to use safety engineered needles, b) poor unsafe work practices (recapping, removal of phlebotomy tube holder), c) failure to dispose properly and d) disposal system failures.

Calculated risk of infection following a needle-stick from an infected source-patient is a requirement for dealing with the occupational exposure to blood borne infectious pathogens. Kane et al reported the risk for hepatitis B virus (HBV) infection, hepatitis C virus (HCV) infection and human immunodeficiency virus (HIV) as 30, 3 and 0.3 % orderly [13]. Due to the very high rate of occupational exposure to blood-borne pathogens, blood-borne pathogens standard was firstly published on December 1991 and effective on March 1992 with the items covering all occupational exposure to blood and other potentially infectious material (OPIM) [12].

According to the standard, determinants of risk of transmission include prevalence of virus in patient population, type of exposure – percutaneous/ muco-cutaneous, extent of injury (superficial or deep), kind of device (hollow bore or solid needle), plasma viremia of source, immune status of health care workers and immediate aftercare as well as application of post exposure prophylaxis. It is accepted that risk is greater in cases with hollow needle, visible blood, deep, large volume and device in artery [12, 14]. Needle stick injuries mainly occur during and after an injection owing to recapping, carrying needles and syringes, patient movement (children) and inappropriate disposal. Basic concepts for prevention of needle stick injuries are a) organizing the physical layout of the Injection work place, b) reducing handling of injection equipment, c)avoid carry, recap or bend, d) cleaning the injection environment prior to and after injections and e) safe disposal to prevent injuries to external public

There are some evidences on incidence of occupational exposure to blood-borne pathogens. Danchaivijitr et al reported that the recalled incidence rate of injuries in the previous 6 months reported by medical workers in Siriraj Hospital was 51.5% [15]. According to this work, recapping and improper disposal of used needles were frequent [15]. Danchaivijitr et al said that it was estimated that 5.9 persons will be HIV infected yearly in Thailand with the same incidence rate of such accidental injuries [15]. Wiwanitkit reported that most of the accidents occurred after venipuncture procedure (80%) [16]. According to the author's view, the incidence of occupational exposure to blood-borne pathogens can be successfully controlled by good work instruction and quality management.

2. Methods of Compliance [17]

2.1. Universal Precautions [18]

Universal precautions must include barriers protection, hand washing, safe techniques, safe handling of sharp items specimens and spill of blood / body fluids, use of disposable / sterile items

2.2. Engineering and Work Practice Controls [19]

Engineering control means controls that can isolate or remove the blood-borne pathogens hazard from the work site. According to CPL 2-2.44D [20], engineering controls will decrease employee exposure either by removing, eliminating, or isolating the hazard. Choosing for engineering and work practice controls is relied on the employer's exposure determination. For exposure determination, the employer have to identify workers exposed to blood or OPIM, review all processes and practices with exposure potential and re-evaluate when new processes or procedures are applied. Also, they must assess available engineering controls (safer medical devices), train employees on safe use and disposal, implement appropriate engineering controls and train employees to use new devices.

2.3. Personal Protective Equipment [21]

There are several new personal protective equipments. Device that do not use a needle for collection of bodily fluids, administration of medication or fluids and any other procedure with potential percutaneous exposure to a contaminated sharp are called "needleless system". The examples of these equipments are phlebotomy needle with "self-blunting" safety feature or "add-on" safety appearance. With personal protective equipment, decreasing of the trend of needle stick injury by more than 70% can be derived [22].

2.4. Management of Occupational Exposure to Blood Borne Pathogens

Post exposure prophylaxis guidelines [23] must include

2.4.1. Immediate First Aid

Immediately cleaning of wound site is the most important part of post exposure prophylaxis (PEP). Skin wounds must be urgently washed with soap and running water. It should be noted that no evidence that antiseptics are useful and caustic agents may do a lot of harm than good. For mucous membranes, flushing thoroughly with clean running water is noted. For eye, irrigation with a liter of saline is recommended for practice.

2.4.2. Report Incident

Sharp injuries record must be systematically done. At least, the log must contain, for each incident: type and brand of device involved, department or area of incident and description of incident.

Some details for PEP of some important blood borne infections will be further noted.

1. HBV [24 – 26]

As previously mentioned, HBV is a high risk for transmission. HBV remains active in dried blood at room temperature for at least 7 days and can be a main cause of transmission. For those health care workers who have been vaccinated, the vaccine provides virtually complete protection to responded vaccinees. Hence, workers should be vaccinated for HBV. However, most are found not vaccinated and 6-10% of vaccinees do not develop antibody. In addition, CDC estimates that 50-75 HCW end their lifes from HBV each year. It also should be noted that HBV risk of disease varies on the HBeAg status. For the clinical hepatitis, both HBeAg and HBsAg positive lead to the risk equal to 22 – 31 % while only HBsAg gives the risk equal to 1 – 6 % [26]. For PEP, immediate vaccination, within 7 days, along with HBIg (0.06 ml/Kg) then complete series of vaccination is suggested for the health care worker without prior history of vaccination or failure of previous immunization by pre-exposure HBV vaccination [26].

2. HCV [25 – 26]

There is no active prophylaxis and immunoglobulins are also not effective. Interferon is also not suggested for prophylaxis. In the absence of prophylaxis against hepatitis C virus (HCV) infection, follow-up management of HCV exposures relies on whether antiviral treatment during the acute phase is selected. Blood test should be performed suddenly and at 6 months for LFT and Anti HCV at 4 – 6 months. After exposure, breast feeding and pregnancy are still permitted for practice.

3. HIV [25, 27]

Considering natural history, HIV infects dendritic cells (DC) then regional lymph nodes before becoming generalized. The risks of transmission in percutaneous exposure and mucous membrane exposure are estimated to 0.3 and 0.09 % [27], orderly. Goal of PEP for HIV is halt viral replication prior to systemic infection is documented. AZT blocks infectivity of HIV infected DC and several animal models showing efficacy. A retrospective study indicates that risk of seroconversion is 81% lower in health care personnel who took AZT for PEP [28]. In addition, peri-natal prophylaxis has been effective. In addition to AZT, reverse transcriptase inhibitors RTI, nucleoside reverse transcriptase inhibitors (NRTI) such as ziduvidine, lamuvidine and protease inhibitors (PI) such as nelfinavir, indinavir are also used for PEP [29]. However, there is no exact supportive evidence comparing single drug versus multiple drugs for PEP.

For HIV PEP, time is considerable. Animal models show that PEP should be given within 2-8 hours of exposure for maximal effect [30 – 31]. PEP may have some benefit up to 1.5 days but will be ineffective if given later. Concerning the cases with seroconversion, it is routinely observed in 6-12 weeks (median time: 46 days). Acute symptomatic sero-conversion can be seem in 50-90% of cases and exposure to symptoms is within average 2-6 weeks [32 – 33]. Therefore, any workers who have a new a flu-like illness in the follow up period should be encouraged to have HIV RNA tested. It is also noted that co-infection with HCV can also delay HIV seroconversion.

References

[1]   Davis YM, McCray E, Simone PM. Hospital infection control practices for tuberculosis. *Clin. Chest. Med.* 1997;18:19-33.

[2]   Tuberculosis Outbreak in a Community Hospital --- District of Columbia, 2002 MMWR Weekly 2004; 53: 214 – 6.

[3]   Ejilemele AA, Ojule AC. Health and safety in clinical laboratories in developing countries: safety considerations. *Niger J. Med.* 2004 Apr-Jun;13(2):182-8.

[4]   Shimosaka H. Collecting blood: its practice, problems, and side effects. *Rinsho Byori.* 2006 Dec;54(12):1223-7.

[5]   Berry AJ. Needle stick and other safety issues. *Anesthesiol. Clin. North America.* 2004 Sep;22(3):493-508, vii.

[6]   Peate I. Occupational exposure of staff to HIV and prophylaxis therapy. *Br. J. Nurs.* 2004 Oct 28-Nov 10;13(19):1146-50.

[7]   Maartens G. Occupational post-exposure HIV prophylaxis. *S. Afr. Med. J.* 2004 Aug;94(8):626-7.

[8]   van Oosterhout JJ, Nyirenda M, Beadsworth MB, Kanyangalika JK, Kumwenda JJ, Zijlstra EE. Challenges in HIV post-exposure prophylaxis for occupational injuries in a large teaching hospital in Malawi. *Trop. Doct.* 2007 Jan;37(1):4-6.

[9]   Wig N. HIV: awareness of management of occupational exposure in health care workers. *Indian J. Med. Sci.* 2003 May;57(5):192-8.

[10]   Kowalska JD, Firlag-Burkacka E, Niezabitowska M, Bakowska E, Ignatowska A, Pulik P, Horban A. Post-exposure prophylaxis of HIV infection in out-patient clinic of hospital for infectious diseases in Warsaw in 2001-2002. *Przegl. Epidemiol.* 2006;60(4):789-94.

[11]   Berry AJ. Needle stick and other safety issues. *Anesthesiol. Clin. North America.* 2004 Sep;22(3):493-508, vii.

[12]   Semes L. The OSHA bloodborne pathogens standard. Implications for optometric practice. *Optom. Vis. Sci.* 1995 May;72(5):296-8.

[13]   Kane A, Lloyd J, Zaffran M, Simonsen L, Kane M. Transmission of hepatitis B, hepatitis C and human immunodeficiency viruses through unsafe injections in the developing world: model-based regional estimates. *Bull. World Health Organ.* 1999;77(10):801-7.

[14]   Purtilo RB. Ethical issues in the handling of bloodborne pathogens: evaluating the Occupational Safety & Health Administration Bloodborne Pathogen Standard. *J. Intraven. Nurs.* 1995 Nov-Dec;18(6 Suppl):S38-42.

[15]   Danchaivijitr S, Kachintorn K, Sangkard K. Needlesticks and cuts with sharp objects in Siriraj Hospital 1992. *J. Med. Assoc. Thai.* 1995 Jul;78 Suppl 2:S108-11.

[16]   Wiwanitkit V. Needle stick injuries during medical training among Thai pre-clinical year medical students of the Faculty of Medicine, Chulalongkorn University. *J. Med. Assoc. Thai.* 2001;84:120-4.

[17]   Cutter J, Jordan S. Uptake of guidelines to avoid and report exposure to blood and body fluids. *J. Adv. Nurs.* 2004 May;46(4):441-52.

[18] Hughes JM. Universal precautions: CDC perspective. *Occup. Med.* 1989;4 Suppl:13-20.

[19] Occupational exposure to bloodborne pathogens; needlestick and other sharps injuries; final rule. Occupational Safety and Health Administration (OSHA), Department of Labor. Final rule; request for comment on the Information Collection (Paperwork) Requirements. *Fed. Regist.* 2001 Jan 18;66(12):5318-25.

[20] U.S. Department of Labor. Occupational Safety & Health Administration. CPL 2-2.44D - Enforcement Procedures for the Occupational Exposure to Bloodborne Pathogens. Standard Number: 1910.1030

[21] Wiwanitkit V. Safety considerations in evacuated blood collection system. *Chula Med. J.* 2000 Aug; 45(8): 643-647.

[22] Whitby RM, McLaws ML. Hollow-bore needlestick injuries in a tertiary teaching hospital: epidemiology, education and engineering. *Med. J. Aust.* 2002 Oct 21;177(8):418-22.

[23] Maynard JE. Preventing transmission of blood-borne pathogens to health care workers. *Natl. Med. J. India.* 2000 Mar-Apr;13(2):82-5.

[24] Hadler SC. Hepatitis B virus infection and health care workers. *Vaccine.* 1990 Mar;8 Suppl:S24-8.

[25] Updated U.S. public health service guidelines for management of occupational exposure to HBV, HCV and HIV and recommendations for post exposure prophylaxis. CDC-MMWR 2001 Jun; 50 (RR-11): 1 – 42.

[26] Puro V, De Carli G, Cicalini S, Soldani F, Balslev U, Begovac J, Boaventura L, Campins Marti M, Hernandez Navarrete MJ, Kammerlander R, Larsen C, Lot F, Lunding S, Marcus U, Payne L, Pereira AA, Thomas T, Ippolito G; The European Occupational Post-Exposure Prophylaxis Study Group. European recommendations for the management of healthcare workers occupationally exposed to hepatitis B virus and hepatitis C virus. *Euro Surveill.* 2005 Oct;10(10):260-4.

[27] Panlilio AL, Cardo DM, Grohskopf LA, Heneine W, Ross CS. Updated U.S. public health service guidelines for management of occupational exposure to HIV and recommendations for post exposure prophylaxis. CDC-MMWR 2005 Sep; 50 (RR-11): 1 – 17.

[28] Cardo DM, Culver DH, Ciesielski CA, Srivastava PU, Marcus R, Abiteboul D, Heptonstall J, Ippolito G, Lot F, McKibben PS, Bell DM. A case-control study of HIV seroconversion in health care workers after percutaneous exposure. Centers for Disease Control and Prevention Needlestick Surveillance Group. *N. Engl. J. Med.* 1997 Nov 20;337(21):1485-90.

[29] Puro V, Cicalini S, De Carli G, Soldani F, Antunes F, Balslev U, Begovac J, Bernasconi E, Boaventura JL, Marti MC, Civljak R, Evans B, Francioli P, Genasi F, Larsen C, Lot F, Lunding S, Marcus U, Pereira AA, Thomas T, Schonwald S, Ippolito G. Post-exposure prophylaxis of HIV infection in healthcare workers: recommendations for the European setting. *Eur. J. Epidemiol.* 2004;19(6):577-84.

[30] Shih CC, Kaneshima H, Rabin L, Namikawa R, Sager P, McGowan J, McCune JM. Postexposure prophylaxis with zidovudine suppresses human immunodeficiency virus

type 1 infection in SCID-hu mice in a time-dependent manner. *J. Infect. Dis.* 1991; 163: 625-7.

[31] Martin LN, Murphey-Corb M, Soike KF, Davison-Fairburn B, Baskin GB. Effects of initiation of 3'-azido,3'-deoxythymidine (zidovudine) treatment at different times after infection of rhesus monkeys with simian immunodeficiency virus. *J. Infect. Dis.* 1993 Oct;168(4):825-35.

[32] Bell DM. Occupational risk of human immunodeficiency virus infection in healthcare workers: an overview. *Am. J. Med.* 1997;102(suppl 5B):9–15.

[33] Ippolito G, Puro V, De Carli G, Italian Study Group on Occupational Risk of HIV Infection. The risk of occupational human immunodeficiency virus in health care workers. *Arch. Int. Med.* 1993;153:1451–8.

Chapter VII

Interesting Aberration of Laboratory Results: Case Studies in Laboratory Medicine

Abstract

Laboratory aberration is often problematic in laboratory medicine. This can be seen in all types of analytical test and should be systematically concerned. This article will present interesting aberration of laboratory results in case studies in laboratory medicine.

Keywords: aberration, laboratory.

Blood Group, Is It Still a Specific Laboratory Result for Each Individual?

1. Introduction

Of several tests in transfusion medicine, blood group testing is the most basic test. Although there are many blood groups, the three most widely used systems are ABO, Rh and MN systems. The clinical usefulness of blood group testing is not limited to transfusion medicine but other purposes [1]. Individual identification is also another widely used indication for blood group testing.

It has been accepted for a long time that blood group is a specific laboratory parameter for each individual. In this article, the author presents to case studies on the changing of blood group parameter in a same individual.

2. Case Studies

Case 1

The medical technologist in-change in the hematology laboratory sent an incident report to the laboratory risk manager on a case of aberrant blood group testing result. She described for a result of blood group of a female patient. She noted that of at that time the patient had blood group "A" , however, this result does not match with the previous report record, blood, group "B" , the laboratory risk manager performed a root cause analysis of this episode and found that the pre-analytical error is the cause. The second blood sample is not the blood from the patient at the first testing. The real patient lent her hospital identification card for her friend and her friend used it for the second testing.

Case 2

A laboratory result of a female pediatric patient with thalassemia after bone marrow transplantation was presented to the hematologist conference. In this case, the blood picture morphology of the patent is normal and there is no anemia . The blood group of the patent is group "A". Her laboratory result before bone marrow transplantation showed hypochromic microcytic anemia 4 + with severe poikilocytosis and her blood group is "O",

3. Discussion

For a long blood group is used as a specific personal identification. It is also used as an important due in personal identification in forensic medicine [1]. Presently, the blood group is also recorded in identification card. In this report a change of blood group testing from an individual is discussed.

The first case is not the actual change. This case shows the importance of pre-analytical error in laboratory medicine [2]. The patient identification is important in specimen collection is necessary [3]. However, in this case, the case is due to the disguising of the patient, which is hard to be detected. The detection in this case is by the comparison with the record in the laboratory database. This implies the usefulness of laboratory information systems. Of interest the episode like this is not an extremely rare incidence. The new question is it is necessary for having the patient's picture in the hospital in identification card.

The second case is an actual evidence of blood group change. This phenomena is due to the transplantation. In real practice, isograft is hardly available. Therefore, the change of some blood antigen corresponding to the new engrafted marrow can be expected. Due to the recent advance on transplantation medicine and stem cell research the change of blood group is possible [4 – 5]. The blood group as a specific identification for an individual is therefore not a totally correct assumption.

A Case Study on Abnormal Low Activated Pential Thromboplastin Result due for Analytical Error

1. Introduction

In hematology the laboratory investigations for coagulation are a important group of hematological test. Prothrombin time (PT) and activated partial thromboplastin time are the two most commonly used coagulation test [6].

Similar to other laboratory investigation aberrant of PT and APTT due to the errors in quality cycle can be seen. Here, the author presents a case study an abnormal low APTT result due to analytical error.

2. Case study

The Clinical pathologist in charge was notified from the medical technologist the abnormal low APTT result. The medical technologist reported that APTT result from the external quality assessment (EQA) program of the laboratory was out of group consensus and about two times lower than results from other laboratory. This incident report was tracing back and it was confirmed that the error was not due to pure analytical or post analytical error but analytical error. The error was a systematic error. A considerable amount of air bubble was detected in the analytical channel of the automated APTT analyzer. However, this error was corrected.

3. Discussion

In medicine, prolonged APTT is on important problem. However, the abnormal low APTT result can be seen. This aberrant low APTT result is usually due to laboratory error. In this article, and interesting case of low APTT result due to a systematic error in analytical phase is presented.

This error can be detected by EQA therefore it is necessary that every laboratory must have EQA [7]. In they case, the air bubble rod in the analyzer can dilute the ratio of blood and reagent in the machine during laboratory analysis and can lead to low abnormal APTT result. Similar phenomena can also be seen in other laboratory analysis especially for arterial blood gas analysis [8].

Problematic Cases in Clinical Chemistry in Laboratory Medicine; Case Studies of Total-Partial Determination

1. Introduction

Laboratory investigation is a basic activity in medicine. In laboratory medicine, determination of substances in blood is the care of laboratory analysis activity. Basically, several principles are used for analysis of substances. However, some substances cannot be directly determined and the others are not easily to perform or very expensive. For those cases, indirect determination is applied. In addition, some tests are reported as proportion for interpretation. For those cases, total and partial determination is needed. Here, the author reports case studies on problematic cases in laboratory medicine according to the total-partial determination.

2. Case studies

Case 1

An aberrant laboratory result of CK and Ck-Mb activities investigation were notified to the physician in charge. The results were CK-MB exceeded CK level. The recheck for the analysis was performed and the result confirmed. The final diagnosis of this case was macro CK phenomenon.

Case 2

An episode of unexpected poor correlation of HbA1C level reported from different laboratories within the hospital was reported to the laboratory. The verification showed real existence of poor correlation. The root cause analysis was performed and the final diagnosis was an error in parameter setting within the automated clinical chemistry analyzer.

3. Discussion

Aberration of laboratory results usually brings confusion for the physicians. Such aberrations can be resulted from errors in laboratory process or real pathological conditions. Here, two interesting case studies on the problematic cases due to real condition. The macro-substance phenomenon is usually common condition leading to unexpected results a presenting with partial component exceeds total component. The macro-substance phenomenon can be seen in many enzymes determination including CK, amylase and transaminase [9 - 10]. In the case, the CK-MB exceed CK level. The cause of macroCK is due to the principle of determination. Fractional immunodetermination of M and B components within CK-MB can lead to problematic case, especially for those with interference immunoglobulins. The electrophoresis for CK isoenzymes or CK isoform can be the clues for correction of the macro CK cases [11 - 12].

Considering the second case, this case is a systemic error in laboratory analysis. The existence of poor correlation can be the first sign of this episode. Indeed, the analysis performed by same analytical principle and type of machine should give similar results [13]. This can also be seen in the good correlation and external quality assurance results. In this case, the observation of the physician can help detect the error or the machine form the manufacturer. Basically, HbA1C determination can be divided into two steps; detection for total Hb and detection for HbA1C; the resulted HbA1C, with % unit, is derived from the factor calculation [14]. Indeed, clinical interpretation of changes in serial measurements of patients' HbA1c ought to be based on the knowledge of pre-analytical, analytical and intra-individual sources of variation that affect the results [13]. In this case, the reverse steps was set within the analyzer, therefor, revered factors were derived. Hence, the abnormal low HbA1C was detected. This type of error should be aware. The effect on following up of diabetic patient can be expected. The good external quality assurance, correlation test between laboratories as will as physician-laboratory communication can be the clue for successful correction of this problem.

Problems with Low Density Lipoprotein Cholesterol (LDL-C) Determination: 2 Case Studies

1. Introduction

Dyslipidemia is a common metabolic disorderwhich can be a contributing factor to coronary heart disease [15]. Lipid profile is a standard laboratory determination in assessing patients with heart conditions [16]. However, determination of total cholesterol and triglyceride alone is not sufficient for clinical assessment, and determination of low density lipoprotein cholesterol (LDL-C), a major lipoprotein relating to coronary heart disease, is recommended. Several methods for LDL-C determination are available, including to beta quantification or ultracentrifugation, lipoprotein electrophoresis, mathematical calculation and direct LDL-C assay [16].

The problems in LDL-C determination are reported and become important problem in laboratory medicine. According to a recent report from Australia, intralaboratory imprecision and sub-optimal interlaboratory comparability are common for LDL-C determination in clinical laboratory [17]. A quality control program for the standardization and harmonization of lipid and lipoprotein analyses is necessary [18]. In addition to quality management, the problems of LDL-determination are sometimes due to the principle of analytical techniques [19]. In this article, the author presented two case studies on the problematic cases of LDL-C determination.

2. Materials and Methods

The author reports two cases studies involving problems with LDL-C determination. These two cases are requested for lipid profile analysis at Department of Laboratory Medicine, Faculty of Medicine, Chulalongkorn University. The first case is a 55 years old male and the second case is a 48 years old female. Both were diagnosed as dyslipidemia. Details of the laboratory investigation results and lipid profiles of the two cases are presented and discussed.

3. Results

Case 1 Friedewald's calculation is a basic calculation technique for determination of LDL-C level in general clinical practice. This technique was firstly described in 1972 [6]. The formula is "LDL-C = TC – (HDL-C) - (TG/5)". In case 1, a questionable result of LDL-C by Friedewald's calculation as LDL-C = -14 mg/dL is discussed. The basic lipid profile in this case was total cholesterol (TC) = 160 mg/dL triglyceride (TG) = 420 mg/dL and high density lipoprotein cholesterol (HDL-C) = 90 mg/dL.

Case 2 Generally, TC consists of two main parts: LDL-C and HDL-C. In case 2, an abnormal laboratory investigation of TC = 259 mg/dL TG = 51 mg/dL HDL-C = 100 mg/dL LDL-C = 172 mg/dL (by direct assay) was returned from the laboratory on one occasion, but the summation of HDL-C-C and LDL-C exceeded TC, which is inconsistent with expected parameters, even abnormal results. Further history taking of the case revealed that previous HDL-C results in this case were abnormal 100 mg/dL.

4. Discussion

It is a basic principle in laboratory medicine that any laboratory determination has limitations, including LDL-C assays. In the first case, a limitation concerning the Friedewald's calculation method for LDL-C determination was exposed, where it becomes unreliable in cases of TG >400 mg/dL [20]. This error is often referred to as a 'train error', since the calculation is based on the previous three parameters; TC, TG and HDL-C determination, and the final error is the summation of errors for each preceding parameter determination. Indeed, concentrations of LDL-C can be monitored by means of the Friedewald formula, which provides a relative estimation of LDL-C concentration when the TG concentration is <200 mg/dl and there are no abnormal lipids. Because of the limitations of the Friedewald calculation, direct methods for an accurate quantification of LDL-C are needed [21]. Although Friedewald's calculation is not a good LDL-C determination method it is still widely used in clinical practice due to its cost effectiveness [22]. However, it may be assumed that during stratification of LDL obtained by calculation the patients are treated too aggressively [23 - 24]. Assuming pharmacological treatment of all mentioned patients, it may be estimated that by using analyses of direct LDL for stratification of probands the costs of hypolipidaemic treatment will by reduced by about 1/4-1/3 [25].

In the second case, a problem with LDL-C determination by direct assay can also be seen. The calculated LDL-C level in the case was equal to 148.8 mg/dL, which is not aberrant, but the summation of LDL-C and HDL-C is more than TC. However the actual cause of interference on LDL-C determination in this case cannot be found. In this case, the proportional of LDL-C components might be abnormal. The ratio of large molecular weight LDL-C, LDL-C-1 and LDL-C-2 might be more than normal and could bring an overestimation of LDL-C [26]. Indeed, increased ratio of the two large molecular weight LDL-C subtypes can be seen in apo E genotype. It is reported that the apo E genotype could exert a significant influence on the estimation of LDL-C concentrations by the Friedewald formula.

5. Conclusion

LDL determination is an important clinical chemistry test. The first case is the problem of Friedewald's calculation method. For solving of this problem, the use of direct LDL assay is recommended [27]. The second case is the possible problem due to increased ratio of the large molecular weight LDL-C. For solving this problem, the use of apo B-based test is recommended [28].

Interference Effect: Case Studies

1. Introduction

At the present, hundreds of clinical chemistry tests are available for general practitioners. These tests are the tool for diagnosis of many diseases. There are a number of requests for clinical chemistry analyses in every medical laboratory daily. For interpretation of the result, the practitioner has to known the basic clinical pathology principle of the requested tests. Sometimes the aberrant and doubtful laboratory results can be detected [29]. In this point, the author presents a case study on the "false high" phenomenon in clinical chemistry results. The main cause of this phenomenon is due to the interference within the blood sample.

2. Cases

The clinical pathologist in charge was consulted for a problem in laboratory result. The case is a male presenting with extremely high triglyceride (1389 mg/dL). The thyroid functiont test of this patients revealed increased FT3 (650 pg/dL), FT4 (2.3 ng/dL) and TSH (6.5 uU/mL). However, the patients has no symptoms of hyperthyroidism. Repeated analysis on fasting sample, all thyroid functions were within normal limit (FT3 520 pg/dL, FT4 1.5 ng/dL and TSH 3.4 uU/mL). The consultation is if there is any error in laboratory procedure.

3. Discussion

Interference is important in any medical laboratory determination since it can affect the laboratory result. The interference can be from the external environment, pathology and physiology of the patient as well as the compositions of the blood sample [29 - 30]. For blood sample, there are three important main compositions affecting the analysis: bilirubin, hemoglobin and triglyceride. Bilirubin causes icteric, hemoglobin causes hemolytic and triglyceride causes lipemic appearances [29 - 30].

Lipemia causes interference because triglyceride concentrations will spuriously increase results for substances when the measurement is based on absorbence of light at the same wavelengths at which lipid particles absorb light. analytes commonly affected are albumin, calcium, and phosphorus but it varies based on machinery. Basically, the inference on the colorimetry tool is usually higher than enzymatic method [29 - 31]. Here, a case of false high or aberrant elevation of thyroid function test is presented. In fact, the patient did not have the hyperthyroidism but the false elevation of thyroid function test is due to the gross lipemia [32 - 33]. Basically, the thyroid function test determination uses the principle of colorimetry by measuring the NADPH reaction. This can be easily interfered by gross lipemia (triglyceride > 1000 mg/dL).

References

[1] Nata M. Forensic medicine and genetic testing. *Nippon Rinsho.* 2005 Dec;63 Suppl 12:395-9.

[2] Howanitz PJ. Errors in laboratory medicine: practical lessons to improve patient safety. *Arch. Pathol. Lab. Med.* 2005 Oct;129(10):1252-61.

[3] Galloway M, Woods R, Whitehead S, Baird G, Stainsby D. An audit of error rates in a UK district hospital transfusion laboratory. *Transfus. Med.* 1999 Sep;9(3):199-203.

[4] Koestner SC, Kappeler A, Schaffner T, Carrel TP, Nydegger UE, Mohacsi P. Histo-blood group type change of the graft from B to O after ABO mismatched heart transplantation. *Lancet.* 2004 May 8;363(9420):1523-5.

[5] Koestner SC, Kappeler A, Schaffner T, Carrel TP, Mohacsi PJ. ABO histo-blood group antigen expression on the graft endothelium long term after ABO-compatible, non-identical heart transplantation. *Xenotransplantation.* 2006 Mar;13(2):166-70.

[6] Sallah S, Kato G. Evaluation of bleeding disorders. A detailed history and laboratory tests provide clues. *Postgrad. Med.* 1998 Apr;103(4):209-10, 215-8.

[7] Tatsumi N, Takubo T, Tsuda I, Hino M. Current problems in quality control (QC) in hematology. *Rinsho Byori.* 1997 Oct;45(10):997-1002.

[8] Gouget B, Manene D, Andrimahatratra R, Bogard M, Gourmelin Y. Pertinence of simultaneous measurements of pO2 and sO2 on ABL 510. *Ann. Biol. Clin.* (Paris). 1992;50(4):247-50.

[9] Maire I, Artur Y, Sanderink GJ. Macroenzymes in human plasma. 1: Macroamylase, macrocreatine kinase, macrolactate dehydrogenase. *Ann. Biol. Clin.* (Paris). 1987;45(3):269-76

[10] Artur Y, Sanderink GJ, Maire I. Macroenzymes in human plasma. 2. Macrogamma-glutamyltransferase, macroalanine aminopeptidase, macroalkaline phosphatase, macroaminotransferases and other macroenzymes. *Ann. Biol. Clin.* (Paris). 1987;45(3):277-84.

[11] Uzawa R. The current aspect of serum creatine kinase in the laboratory medicine. *Rinsho Byori.* 1990 Mar;38(3):282-7.

[12] Kanemitsu F, Okigaki T. Creatine kinase isoenzymes. *J. Chromatogr.* 1988 Jul 29;429:399-417.

[13] Matteucci E, Cinapri V, Rossi L, Lucchetti A, Giampietro O. Glycated hemoglobin measurement: intermethod comparison. *Diabetes Nutr. Metab.* 2001 Aug;14(4):217-9.

[14] Charuruks N, Milintagas A, Watanaboonyoungcharoen P, Ariyaboonsiri C. Determination of reference intervals of HbA1C (DCCT/NGSP) and HbA1C (IFCC) in adults. *J. Med. Assoc. Thai.* 2005 Jun;88(6):810-6.

[15] Sakurabayashi I. Hyperlipemia and nutritional examination. *Rinsho Byori.* 2003 Oct;Suppl 127:79-85.

[16] Kubo N. Evaluation of LDL-C-cholesterol measurement. *Rinsho Byori.* 2002;50:242-7.

[17] Chennell A, Sullivan DR, Penberthy LA, Hensley WJ. Comparability of lipoprotein measurements, total:HDL cholesterol ratio and other coronary risk functions within and between laboratories in Australia. *Pathology.* 1994;26:471-6.

[18] McGuinness C, Seccombe DW, Frohlich JJ, Ehnholm C, Sundvall J, Steiner G. Laboratory standardization of a large international clinical trial: the DAIS experience. DAIS Project Group. Diabetes Atherosclerosis Intervention Study. *Clin. Biochem.* 2000;33:15-24.

[19] Wiwanitkit V. Total and partial determination in biochemistry laboratory tests. *Chula Med. J.* 2000; 43: 109-113.

[20] Friedewald WT, Levy RI, Fredrickson DS. Estimation of the concentration of lowdensity lipoprotein cholesterol in plasma, without use of the preparative ultracentrifuge. *Clin. Chem.* 18:499–502, 1972.

[21] Chotkowska E, Kurjata P, Kupsc W. Evaluation of the precision of the Friedewald's formula for the calculation of low density lipoprotein cholesterol concentration in serum. *Pol. Merkuriusz Lek.* 2001;11:348-51.

[22] Turkalp I, Cil Z, Ozkazanc D. Analytical performance of a direct assay for LDL-cholesterol: a comparative assessment versus Friedewald's formula. *Anadolu Kardiyol. Derg.* 2005;5:13-7.

[23] Asswawitoontip S, Wiwanitkit V. Cost-effective study of determination methods for low-density lipoprotein by new direct assay compared to Friedewald's formula calculation in hypercholesterolemic subjects. *J. Med. Assoc. Thai.* 2002 Jun;85 Suppl 1:S91-6.

[24] Stejskal D, Pastorkova R, Frankova M, Bartek J, Horalik D. Benefit of direct determination of LDL-cholesterol (comparison with LDL measurement using calculated estimates. *Vnitr. Lek.* 1998;44:707-13.

[25] Cole TG, Ferguson CA, Gibson DW, Nowatzke WL. Optimization of beta-quantification methods for high-throughput applications. *Clin. Chem.* 2001;47:712-21.

[26] Tremblay AJ, Bergeron J, Gagne JM, Gagne C, Couture P. Influence of apolipoprotein E genotype on the reliability of the Friedewald formula in the estimation of low-density lipoprotein cholesterol concentrations. *Metabolism.* 2005;54:1014-9.

[27] Smets EM, Pequeriaux NC, Blaton V, Goldschmidt HM. Analytical performance of a direct assay for LDL-cholesterol. Clin. *Chem. Lab. Med.* 2001;39:270-80.

[28] Bairaktari E, Hatzidimou K, Tzallas C, Vini M, Katsaraki A, Tselepis A, Elisaf M, Tsolas O. Estimation of LDL cholesterol based on the Friedewald formula and on apo B levels. *Clin. Biochem.* 2000;33:549-55.

[29] Wiwanitkit V. 0 Interesting cases in aberrant biochemistry laboratory results. *Chula Med. J.* 2002 Dec; \cf0\f0 46(12) : 997-1001

[30] Yamazaki H, Nanbu S. Recent advances in biochemistry in clinical pathology. *Rinsho Byori.* 1971 Feb;19(2):91-9.

[31] Westgard GO. Precision and accuracy: concepts and assessment by method evaluation testing. *Crit. Rev. Clin. Lab. Sci.* 1981;13(4):283-330.

[32] Keffer JH. Preanalytical considerations in testing thyroid function. *Clin. Chem.* 1996 Jan;42(1):125-34.

[33] Andersen S, Bruun NH, Pedersen KM, Laurberg P. Biologic variation is important for interpretation of thyroid function tests. *Thyroid.* 2003 Nov;13(11):1069-78.

Chapter VIII

Automation in Laboratory: What Are New Things?

Abstract

Automation in laboratory is the big jump in laboratory medicine. Because the classical manual method of laboratory analysis needs a lot of medical personnel for performing of the test and this can be time-consuming and give a slow turnaround time. Therefore, there must be a new thing to solve this problem. The necessary technology must bring fast analysis and automation is the solution for this query. The automation has been introduced to the medical laboratory for years. In this article, the author will briefly discuss and detail on some important new things in automation in medical laboratory.

Keywords: automation, laboratory.

Introduction

Automation in laboratory is the big jump in laboratory medicine. Because the classical manual method of laboratory analysis needs a lot of medical personnel for performing of the test and this can be time-consuming and give a slow turnaround time. Therefore, there must be a new thing to solve this problem. The necessary technology must bring fast analysis and automation is the solution for this query. The automation has been introduced to the medical laboratory for years. Until present, there is still continuous research and development on this area. In this article, the author will briefly discuss and detail on some important new things in automation in medical laboratory.

Autmation in Hematology: From Classical Technology to Flow Cytometry

The automation in hematology is the first group of automation in laboratory medicine. Because hematological test, especially for complete blood count (CBC) is the widely used clinical laboratory test for a long time. This test bases on the microscopic examination of blood smear and clinical chemistry manual test for hemoglobin. This takes a long period of time for each analysis. As previously said hemaological test is frequently performed test that has limitation in manual patice [1]. It is also influenced by physiologic variance, interpretive bias, sampling errors, and the non-random distribution of cells in a blood smear [1]. The result is data that can be statistically imprecise, insensitive, and nonspecific [1]. Therefore, there is a need to find new automatic test. Luckily, the automated hematology analyzer was finally developed. The first machine bases on the principle of electrical impedance for cell counting which is further developed into newer technology, flow cytometer. Accompanied with staining technique, several staining methods are implemented in the automated clinical hematology analyzer and this bring the new jump in hematology. Charuruk said that automated assays were more precise than their manual counterparts, and frequently are assumed to have greater value and many new hematologic parameters could be generated by these analysers [1]. Charuruks et al reported that the automated hematocrit was a rectilinear function of the erythrocyte count while the manual hematocrit value deviated systematically from that of the automated hematocrit, its correlation with the erythrocyte count was more curvilinear [2]. In addition, they noted that the automated MCVs were nearly constant throughout the range of dilutions of erythrocytes and showed that hematocrit and MCV by the automated method was more reliable than by the manual method [2]. With the advanced in flow cytometry, the automated hematology analyzer can be used for analysis of reticulocyte, immature red blood cell, immature platelet as well as lymphocyte subset at present.

The automatic technique can bring diminishing labour tasks and to significant improvements in accuracy and precision compared with the manual microscopic methods, however, it also adds a considerable expense to the laboratory budget that needs further thought for coping with [3]. Charuruks et al successfully developed new technique to lowering the cost of analysis [3]. For example, Charuruks et al reported the modified method of applying the new mixture of 1 uL of whole blood with 1 mL of reticulocyte reagent, which they evaluated for its accuracy and precision, instead of using the mixture of 3 uL of whole blood with 3 mL of reticulocyte reagent recommended by the company [3]. In this work, Charuruks et al demonstrated the accepted accurate and precise results of percentage and absolute number of retculocyte count, low-stained reticulocyte count and its corpuscular indices; the mean reticulocyte corpuscular volume (MCVr), mean reticulocyte corpuscular hemoglobin concentration (CHCMr), and mean reticulocyte hemoglobin content (CHr) [3]. Finally, Charuruks et al suggested that, for every red cell assessed, the number, the cell volume, hemoglobin content and concentration are accurately and precisely measured by the modified method while the sub-populations of reticulocyte count and distribution width of reticulocyte indices were variable [3].

Electrochemiluminescence: A New Analytical Mean in Clinical Chemistry and Immunology [4]

There are several means for analysis in clinical chemistry and immunology. However, an important new modern technique is based on the electrochemistry principle. Electrochemiluminescence or electrogenerated chemiluminescence(ECL) is a specific form of chemiluminescence (CL) in which the light emitting chemiluminescent reaction is preceded by an electrochemical reaction [5 - 6]. The advantages of CL are confirmed. The electrochemical reaction permits the time and position of the light emitting reaction to be monitored. By controlling the time light emission can be delayed until events such as immune or enzyme catalyzed reactions have already occurred [5 - 6]. For mechanism, the combination of ECL with universally coated micro particle technologies leads to very fast assays with very high sensitivity and broad measuring range [5 - 6]. This is basically set by optimised selection of the electrochemically active "substrates" (Ruthenium chelates and tris-propylamine, TPA) [7]. Firstly, the substrate will reach the cathode to recieve in the excessive electron and turns into negative charge. Secondly, the analyte will go to the anode to spit out electron and becomes positive charge. Thirdly, substrate and analyte interact each other. The transmission of excessive electron from substrate to the electron-depleted analyte happened. The electron transmission is based on the scientific rule of electron valency change. Finally, the transmission of electron causes the transmission of photon that can be recorded and further interpreted as specific corresponding amount of analyte. The use of a confocal microspectrometer permits in situ photon detection of the ECL reaction with micrometric spatial resolution and detects the concentration distribution of the photon source in the vicinity of specific interelectrode gap [8]. ECL is widely used as an advanced technique for new clinical chemistry analyzer for microanalyte including cardiac markers, tumor markers and hormones [9].

What Is the Future Trend of Automation in Medical Laboratory [10]

Health care system was out of economic control in many countries owing to recent world economiocal problem. Medical personnel usually spend a higher percentage of gross national product for health care than before with no sign of improvement. Charuruk said that "Entering the twenty-first century the pressure from globalization to standardize medical care quality, to extend medical care to a greater percentage of the population and the need to control costs represent the greatest challenges to health care system reform [10]". To cope with a huge amount of laboratory samples and consequent labor problems within the medical laboratory, the emerging concepts of Laboratory Automation Systems (LAS) and Total Laboratory Automation (TLA) can bring a vision of how the central laboratory in the future might walk [10]. As a consequence, health care providers must consider the need to complement centralized laboratory facilities with satellite laboratories or a Point-of-Care Testing (POCT) network that provide more rapid results [10]. To prepare and plan for what will occur in the future for automation or big laboratory, a plan for automation laboratory management is needed.

1. Total Laboratory Automation

TLA is the concept to consolidate all instruments in all phases of the laboratory into the single system. This can help unnecessary division and helps decreased unnecessary step. This TLA can cover the pre-analytical phase by introduction of pre-analytical phase machine that can help decrease the pre-analytical error in laboratory analysis [11]. Selecting automation for the medical laboratory in the present day tend to the selection of the total automation system [12]. The future plan for each medical laboratory should bend its direction to some degrees of TLA [13]. At least, consolidation should be performed [14].

2. Point-of-Care Testing

POCT is another important concept. This hits to the point that the patient is at ward and the fastest analysis must be at ward. This is originated from small simple capillary blood analysis at bed side. This concept becomes widely used in laboratory medicine at present. There are also many reports in this area. There are also many reports on evaluation of the new POCT tools. Here, the author will present example on such evaluation in the model of POCT for lipid assay. Basically, lipid is one of the important nutrition for human. Similar to other nutrition, the defect of lipid metabolism can be seen in medicine. To determine the lipid level in medicine, clinical chemistry technique is used. The lipid profile is a group of tests that are often request together to determine risk of coronary heart disease [15]. The tests that make up a lipid profile are tests that have been demonstrated to be good indicators of whether a patient is likely to have an obstruction of blood vessels [15]. Some of the genetic factors and environmental factors such as diet and drug therapies that are clarified to increase the risk of hyperlipidemia and possibly the predisposition to cardiovascular disease are documented in medicine [16]. The cholesterol associated with the low-density lipoproteins (LDL), accounting for 60-75% of the plasma levels, is corresponding for the powerful and direct relationship which exists between plasma cholesterol and coronary heart disease [16]. Also, the cholesterol that accumulates in atheromatous pathological sites is derived primarily from LDL in plasma [16].

With the modern concept of point-of care testing, the monitoring of lipid profile by bed-side device is available. Walk-in screening programs are becoming popular as a mean of health care delivery. Self-monitoring of plasma triglycerides and total cholesterol can be a very useful tool to monitor the lipid, on a daily basis. Many new devices have been launched for the bed-side monitoring of lipid profile. Some devices such as Accutrend GCT are accompanied with the determination of glucose. Because it uses capillary blood and gives results in minutes, the system is ideally and basically designed to be used outside the laboratory. The machine is also suited for home use by the hyperlipidemic patients under medical supervision. Here, the author perform a cost identification to compare between the determination for total cholesterol and triglyceride by bed –side meters comparing to standard clinical chemistry analyzer in the view of patient or customer. In this small study, as described, the two studied alternatives in this work is a) bed-side meter, Accutrend GCT (Roche) and b) standard clinical chemistry analyzer (Hitachi) The primary assumption is that the efficacy of both alternatives is the same. Cost analysis was further done. The cost in baht (1 US dollar = 41 baht) for performing each test was reviewed. The cost used was set as the price of each test at the reference laboratory in Thailand (Division of Parasitology, King Chulalongkorn Memorial Hospital, Bangkok Thailand). Then cost comparison was performed._The cost for each alternative node for total cholesterol and triglyceride determination is calculated. The cost identification analysis is then performed. The author also determines for the equilibrium point between both alternatives. Concerning the results, cost of each alternative method for total cholesterol and triglyceride determination are presented in Table 1. The equilibrium between both alternatives is at 92 tests.

Lipid disorders is at high rank among the risk factors known to be associated with development of cardiovascular diseases which constitute a great socio-economic problem in many countries [17]. Blood lipid lipoprotein and apolipoprotein determinations belong in all types of medical services to the most important ones not only because of their clinical diagnostic but also of their clinical predictive value [17]. The lipid profile includes total cholesterol, HDL-cholesterol , LDL-cholesterol , and triglycerides. Sometimes the report will cover an additional calculated values such as HDL/Cholesterol ratio or a risk score based on lipid profile results, age, sex, and other risk factors. However, the two routine recommended tests are total cholesterol and triglyceride [18].

Here, the author performed an economical analysis to compare cost for determination of total cholesterol and triglyceride for the patient. Here, it can be shown that the cost per test for bed-side meter is more expensive than standard test if the number of the test is less than 92 tests. It means that if the patients buy the bed side meter, it must take about 8.6 years use to reach the equilibrium (if the lipid profile follow up is recommended as monthly follow up). Therefore, the bed side meter is not a good test for a short term usage, based on medical laboratory economics principles.

Table 1. Cost identification for each method for determination for total cholesterol and triglyceride in the patient view

	Bed side meter		Clinical chemistry analyzer	
	item	Cost (baht)	item	Cost (baht)
Fix cost	machine	6,900	N/A	0
Variable cost*	Strip	120	Test charge	200
	Specimen collection	5		

*varied due to the number of the test
** equilibrium (N) can be calculated as "6900 + [(120 + 5) x N] = (200 x N)".

Monitoring of Lipid Profile
by Bed-Side Device

As already mentioned for the effectiveness and economical aspect on the lipid measurement by POCT the author might feel interested on this kind on machine. In order to help the reader better understand on the POCT, there author hereby details on the monitoring of lipid profile by bed-side device as a model in laboratory medicine.

1. Introduction to Lipid Profile Monitoring

Hyperlipidemia is defined by increased levels of the plasma levels; the risk for atherosclerosis is relating to the classification types IIa, IIb, III, and possibly IV, a classification system based on phenotypic manifestations of increased lipoprotein fractions [16]. In addition, low-grade inflammation, enhanced oxidant stress and lipid peroxidation have been demonstrated for the specific association with increased cardiovascular risk associated with cardiovascular events [17]. It has been hypothesized and proposed that the low-grade inflammatory state characterizing metabolic disorders such as obesity, hypercholesterolemia, type 2 diabetes mellitus and homozygous homocystinuria might be the primary trigger of thromboxane-dependent platelet activation mediated through enhanced lipid peroxidation [17]. The Lipid Research Clinics Program reports clinical information on plasma lipid and lipoprotein cholesterol distributions of a large-scale screening of white men and women aged 20-59 years in the US [17]. They found age-related trends for rising triglycerides and cholesterol with differences between the sexes clearly demonstrated [16]. Lipid profile is recommended that lipid profile should be included in a routine annual check up program [19]. According to a recent report in Thailand, pathological values of LDL-cholesterol were determined in 26.3% of the males and 28.4% of the females [20].

2. Measurement of Lipid Profile in Clinical Laboratory

Lipid disorders rank high among the risk factors known to be associated with development of cardiovascular diseases which constitute a great socio-economic problem in many countries [21]. Blood lipid lipoprotein and apolipoprotein measurements belong in all types of medical services to the most important ones not only because of their diagnostic but also of predictive value [21]. The lipid profile includes total cholesterol, HDL-cholesterol , LDL-cholesterol , and triglycerides. Sometimes the report will include additional calculated values such as HDL/Cholesterol ratio or a risk score based on lipid profile results, age, sex, and other risk factors. However, the two routine recommended tests are total cholesterol and triglyceride [17].

Presently, lipid profile is usually determined by automated clinical chemistry analyzer, based on the principle of colorimetry and spectrometry. Twelve-hour fasting is always preferable for blood cholesterol testing. Fasting means no food or drink.

For HDL-cholesterol and LDL-cholesterol, the direct determination by automated clinical chemistry analyzer are also available at presented. In addition, the lipoprotein-cholesterol electrophoresis is also available for the fractional study of dyslipidemia [22].

3. Monitoring of Lipid Profile by Bed-Side Device

With the new concept of point-of care testing, the monitoring of lipid profile by bed-side device is available. Walk-in screening programs are becoming popular as a method of health care delivery. Self-monitoring of plasma triglycerides and total cholesterol may be a very useful tool to monitor, on a daily basis. Many new devices have been launched for the bed-side monitoring of lipid profile. Some devices such as Accutrend GCT are accompanied with the determination of glucose. Because it uses capillary blood and gives results in minutes, the system is ideally produced to be used outside the laboratory. The machine is also proper for home use of patients under medical supervision. The system usually has two parts, the meter and the strip. Basically, the meter measures the intensity of the reaction color within the reaction layer of the test strip by reflectance photometry and further determines the parameter concentration of the sample through a lot-specific algorithm calculation. The result is displayed in mg/dL or mmol/L and stored automatically in the analyzer with time and date. The resulted parameters can be derived within a few minutes (3 – 5 minutes). Considering the strip, it contain four main parts as a) a protective mesh, impregnated with a specific surfactant, b) a glass fiber fleece that acts as a separating layer for blood cells, c) a reaction film in which the color formation happen and c) a bar code on the reverse side, which is determined by the meter to confirm test strip identity. The test strip should be kept at room temperature (2 - 30 °C).

The system has been tested for its effectiveness for a few recent years. The results from the comparative studies indicated that the determined triglyceride and total cholesterol results were concordant with those from clinical chemistry analyzers [23].

Also, the meter is used in many studies as a program from lipidemia pharmacotherapy [24 – 25]. For example, the program at the University of California-Irvine Medical Center administers POCT, low-cost lipid profile testing, directly dealed with cases in their own care, and gave individualized education to patients regarding cardiovascular risk reduction [24].

Biochip: New Generation of Automation

1. Introduction

With the start of the development of tiny chip chemical sensors in the past two decades, which can routinely be mounted in the tip of a catheter, it was accepted that clinical chemistry would come to the new phase: on site service. Continuous in vivo monitoring of many important blood analytes is expected to replace the old technique with off-line laboratory analyses. The fusion of microelectronics and molecular biology has created a modern technology, which gives a rapid, efficient, and cost-effective tool in molecular diagnostics at a high-sample throughput [26]. The biochip has recently been chose as one of the ten scientific highlights in the year 1998 [26]. The application of microelectronics can be from the polymerase chain reaction (PCR), nucleotide sequence analysis via DNA-chips or capillary electrophoresis-chips to gene expression analysis [26]. In the biochip, flow splitting microchannels, chaotic micromixers, reaction microchambers and detection microfilters are all completely integrated. Several biochips are accepted as an important role in molecular diagnostics, and their application in POCT diagnosis is expected to facilitate the development of new concept of personalized medicine [27]. In this article, the summarization on the developmental process and important new types of biochips is hereby made and presented.

2. Developmental Process of a Biochip

2.1. Fabrication

Fabrication step is a significant factor and prompt the first consideration in biochip design. The fabrication of this memory system relies on the self-assembly of the nucleic acid junction system, which performs itself as the scaffolding for a molecular wire consisting of polyacetylene-like units [28]. Microstructures on polymer substrate have been constructed and assembled by making use of simple several technologies including to laser-guided direct writing, injection moulding-thermal bonding and hot-embossing-UV adhesive bonding. Laser-induced optical forces can be applied to lead and deposit 100 nm - 10 microm-diameter particles onto solid surfaces in a process naemly "laser-guided direct writing" [29]. Nearly any particulate material, including both biological and electronic materials, can be done and deposited on surfaces with micrometer accuracy [29]. The microinjection moulding of copolymer and thermal bonding of injection moulded polymer substrates can be used in cases of mass production [30]. Passive elements can be hot embossed into the polymer microchips and then, the chips can be collected altogether into a three-dimensional architecture with the interconnect fabricated from an elastomer to create a leak-free connection between the biochips [31]. The array can be then produced making use of a photomodification process, which involved three steps; (a) UV exposure of the polymer surface; (b) coupling and (c) washing of the surface [31]. A molecular switch to control current is documented and this is based on the formation of a charge-transfer complex [3]. A molecular-scale bit is presented which follows the concept on oxidation-reduction potentials of metal atoms or clusters [29]. The readable bit which can be constructed of these components has a specific volume of 3 x 10(7) A3, and should operate at electronic speeds over short distances [29].

2.2. Microfluidic Control

The control of microfluidics in a device has been one of the most focused interested issues. Among all kinds of body fluidics, the blood, which runs the beneficial ingredients to individual organs and carries away the by-products, has abundant health-related information and hence becomes the most interested target in clinical diagnostic applications. The flow principle is mentioned as a valveless, electroosmotically driven technology applied for controlling the stream profile in a laminar flow chamber [32]. Basically, a blood sample has to pass three specific necessary steps within a biochip: (a) sample loading; (b) flowing for analysis and (c) explusion of already analyzed sample. Adjustment of the flow of electroosmotically controlled guiding stream, running parallel to a central sample stream, can be applied for positioning the sample stream in the dimension perpendicular to the flow direction [32]. The analyte sample can be loaded from the sample inlet port to the detection chamber by capillary force, with no any external intervening forces [33]. For this and to control the time duration of sample fluid in each specific part of the device, including the inlet port, diffusion barrier, reaction chamber, flow-delay neck, and detection chamber, the fluid conduit has been designed with various geometries of channel width, depth, and shape [33]. Specifically, the fluid path has been set so that the sample flow naturally stops after filling the detection chamber to let sufficient time for biochemical reaction and subsequent washing steps [33]. The feasibility of creating nanometer scale depressions in biological substrates making use of active enzymes delivered with scanning probe microscopes has been previously demonstratedand and provided a simple mean to precisely control of fluid flow dimension [34]. Such nanochannels may be successfully used in nanofluidic biochip applications [34].

2.3. Amperometric Measurement

The basic concept of the on-chip biosensor array is based on amperometric measurements. Tranmembrane flux monitoring is used and further implied as laboratory results. Basically, flux change due to the biochemical reaction is temporally and can be amperometrically determined by biochip microelectrodes [35]. Based on electrical biochips made in Si-technology portable devices have been constructed for field applications and point of care diagnosis [36]. These miniaturized amperometric biosensor devices let the evaluation of biomolecular interactions by determining the redox recycling of ELISA products, as well as the electrical monitoring of specific metabolites [36]. The highly sensitive redox recycling is helped analysis by interdigitated ultramicroelectrodes of high spatial resolution [36].

3. Important New Types of Biochips

3.1. GeneChip

GeneChip is a new tool developed by Affymetrix USA. It is based on the concept of a specific mean for spotting DNA probes on chips, which is different from any other DNA chips, and can end up the whole process from sample preparation to information construction and analysis [37]. The GeneChip system can be used for both gene expression analysis and genomic mutation analysis, which would play a significant role in human genome analysis in the future [37]. DNA microarray technology, especially the use of GeneChip microarrays, has turned into a standard tool for parallel gene expression analysis [38]. Recent improvements in GeneChip microarrays helps create whole-genome expression analysis [38]. GeneChips can be applied for high-throughput mutation detection, single nucleotide polymorphism genotyping, expression profiling and detection of chromosomal aberrations [39]. This in turn reveals the way for clinical applications in genetics, cytogenetics, pharmacogenetics, oncology and pathogen recognition [40]. Establishing standards is a core issue in improving data quality and, in combination with automatic, easy-to-interpret reports, will construct the basis of the clinical applicability [40].

3.2. CYP450

Pharmacogenetics has accepted increasing importance with the developing concepts of personalized medicine. There is a requirement to determine the metabolic status of an individual when applying drugs, the actions of which are influenced by drug-metabolizing enzymes. CYP450 is an enzyme noted for its specific correlation to pharmacodynamics of many drugs. Laboratory determination for CYP450 presently comes to its advanced phase, biochip technology. The most well know chip for CYP450 is AmpliChip CYP450 [41]. The AmpliChip CYP450 Test is made based on microarray technology, which combines hybridization in precise locations on a glass microarray and a fluorescent labeling system [42]. AmpliChip CYP450 (Roche Molecular Diagnostics, Alameda, CA, USA) is the first approved microarray molecular diagnostic tool for the determination of 29 polymorphisms and mutations of the CYP2D6 gene, and two polymorphisms of the CYP2C19 gene [42]. It makes use of both Roche's PCR technology with the GeneChip microarray system (Affymetrix, Santa Clara, CA, USA) [42]. The role of CYP450 chip for diagnosis for CYP polymorphisms in the treatment use of many drugs can be expected [42].

3.3. Protein Biochips

Old methods of protein analysis such as electrophoresis, ELISA and liquid chromatography are routinely time-consuming, labor-intensive and lack high-throughput capacity [43]. In addition, all existing means used to measure proteins necessitate division of the original sample and individual tests brought out for each substance, with an associated cost for each test [43]. Proteomics' requirement for simultaneous determination of multiple markers is now possible with biochip array technology [44]. Several laboratories utilise in-house, manual protocols for biochip fabrication and sample testing [44]. Reproducibility and standardisation of biochip processes is vital to confirm quality of results and give the best tool for elucidation of complex relationships between multiple proteins in diseased conditions [45]. The chip system lets several tests to be performed simultaneously without classifying the original patient sample [44]. This system facilitates the development of multiplexed assays that simultaneously determine several different analytes in a small sample volume [44]. These emerging technologies can be divided into two categories: 1) spotted array-based tools, and 2) microfluidic-based tools [44].

3.4. Nanotechnology-Based Biochips

There are several attempts to create new biochips based on new nanotechnology. The resonance-enhanced absorption (REA) by metal clusters on a surface is a new technique on which to base bio-optical devices [46]. A four-layer device having a metal mirror, a polymer or glass-type distance layer, a biomolecule interaction layer and a sub-monolayer of biorecognitively bound metal nano-clusters is done and used for biochips [46]. In addition, polymer microlenses have been produced by delivering droplets of a monomer mixture to a glass substrate maing use of a nano fountain pen (NFP) [47]. Subsequent UV polymerization yielded microlenses with optical microparts that were controlled by varying the deposition time of the monomer solution [47]. Making use of this approach, it is probable to strategically place single microlenses at predefined positions with very high accuracy, an ability which can prove very useful for nano-biochip applications [21]. Accompanied with nanolithography technique, nanochannel biochip might be effectively fabricated by NFP [9, 22].

Opinion Survey of Targeted Physicians for Using of New Urinalysis Service by Automated Analyzer, IQ-200: An Example

1. Introduction

Laboratory service can be considered as service that bases on client theory. The main clients of medical laboratory are a) patients and b) physicians. To serve the clients is the goal of service. According to the hospital accreditation principle, satisfaction management is important [48 - 49]. Usually, the laboratory focuses to make the patients the most satisfaction. However, only a few concerns are given to another client groups, the physician.

Basically, client survey is a basic research for service marketing. This concept can be modified for laboratory service [50 - 51]. How, the author reports the opining survey of targeted physicians for using of new laboratory technique for urinalysis.

2. Materials and Methods

A. New test

A new test is expected to be in a real service in a tertiary hospital. How to know the user reflection to this test is necessary and should be systematically done. The test is automated urinalysis. A new automated urinalysis analyzer, namely IQ-200 [52 – 53] is proposed for usage in routine rule. There are some reports mentioning for its good efficacy and comparable result to classical urinalysis.

B. Opinion survey

The author performed on opinion survey on the usage of $\pm$ Q-200. First, the laboratory distributed information on analytic activity of IQ – 200 to focus targeted group, physicians in nephrology unit of Department of medicine and pediatrics and in Urology unit of surgery. Accompanied with the information, the laboratory asked the corresponding physicians in all units to give any opinions or comments on the use of IQ-200 in real practice by returning the information to the laboratory within due date.

3. Results

Only the physicians form nephrology unit of pediatrics gave information back within due date. There was a comment that IQ-200 is believable for urinalysis only in case of normal urine sediment. Based on the comment, the laboratory decided to use IQ-200 for screening purpose.

4. Discussion

Evaluation of a new analyzer before real usage is necessary Basically both analylical and clinical diagnostic property must be evaluated. However there is a fact to notify that a success in service must consider the acceptance of the client. The opinion survey of the physicians as clients is a new idea [54]. Basically the laboratory usually launches new test into use without communication with the targeted physicians. Sometimes, the physicians are unfamiliar to the new techniques therefore they did not use or misuse the new test. Given the information can help solve there problem. In addition, two - way communication, listening to the physicians opinions can help manage the planning for service of the new test; proper setting, proper tine for serve and amount of analyzer [55].

In deed, the determination of response of targeted clients before launching of new laboratory program can help determine the success of the program. Recently, Charuruks et al performed a similar questionnaire survey to get the opinion of the targeted system for blood gas analyzer [56]. This type of study is recommended to co-operate the success in management of new laboratory test. Expert opinion is at least a low grade data in evidence-based medicine.

References

[1] Charuruks N. Technicon H : an automatic hematology analyser in medicine. *Chula Med. J.* 1995 ; 40(7) : 585-600.

[2] Charuruks N. Krailadsiri P, Punyadilok S. Comparison of erythrocytic indices by automated and manual methods. *Chula Med. J.* 1993; 37 (5) : 317-326.

[3] Charuruks N, Huayhongthong V, Srisink N, Yangsuk S, Dangchen N, Pirom S, Thaisamsen M, Noysri P. Evaluation of the reticulocyte count portion of technicon H*3 blood analyzer : lowering the test expense by reducing the reticulocyte reagent. *J. Med. Assoc. Thai.* 1997; 80(Suppl 1): S62-S71

[4] Ritcher NM. Electrochemiluminescence (ECL). Chem Rev. 2004 Jun;104(6):3003-36.

[5] Kissinger PT. Electrochemical detection in bioanalysis. *J. Pharm. Biomed. Anal.* 1996;14:871-80.

[6] Richter MM. Electrochemiluminescence (ECL). *Chem. Rev.* 2004;104:3003-36.

[7] Bruce D, McCall J, Richter MM. Effects of electron withdrawing and donating groups on the efficiency of tris(2,2'-bipyridyl)ruthenium(II)/tri-n-propylamine electro chemi luminescence. *Analyst.* 2002;127:125-8.

[8] Amatore C, Pebay C, Servant L, Sojic N, Szunerits S, Thouin L. Mapping electrochemiluminescence as generated at double-band microelectrodes by confocal microscopy under steady state. *Chemphyschem.* 2006;7:1322-7.

[9] Wiwanitkit V. Advanced technique for sex hormone determination by electrochemiluminescence and application to disability. *Sexual Disabl.* 2007; 25(3): 141 - 6.

[10] Charuruks N. Future trend of automation laboratory management in Thailand. *Chula Med. J.* 2002; 46(4) : 289-302.

[11] Hawker CD. Laboratory automation: total and subtotal. *Clin. Lab. Med.* 2007 Dec;27(4):749-70, vi.

[12] Mealanson SE, Lindeman NI, Jarolim P. Selecting automation for the clinical chemistry laboratory. *Arch. Pathol. Lab. Med.* 2007 Jul;131(7):1063-9.

[13] Kawagoe I, Kanno T. Dawning of laboratory automation; individually constructed automated systems. *Rinsho Byori.* 2000 Oct;(Suppl 114):1-7.

[14] Seaberg RS, Stallone RO, Statland BE. The role of total laboratory automation in a consolidated laboratory network. Clin Chem. 2000 May;46(5):751-6.

[15] Kannel WB. Lipid profile and the potential coronary victim. *Am. J. Clin. Nutr.* 1971 Sep;24(9):1074-81.

[16] Nestruck AC, Davignon J. Risks for hyperlipidemia. *Cardiol. Clin.* 1986 Feb;4(1):47-56.

[17] Grafnetter D. International quality assurance schemes for cholesterol. *Scand. J. Clin. Lab. Invest. Suppl.* 1990;198:32-42.

[18] Davi G, Falco A. Oxidant stress, inflammation and atherogenesis. *Lupus.* 2005;14(9):760-4.

[19] Wiwanitkit V. Abnormal laboratory results as presentation in screening test. *Chula Med. J.* 1998; 42:1059-67.

[20] Phonrat B, Pongpaew P, Tungtrongchitr R, Horsawat V, Supanaranond W, Vutikes S, Vudhivai N, Schelp FP. Risk factors for chronic diseases among road sweepers in Bangkok. *Southeast Asian J. Trop. Med. Public Health.* 1997 Mar;28(1):36-45.

[21] Grafnetter D. International quality assurance schemes for cholesterol. *Scand. J. Clin. Lab. Invest. Suppl.* 1990;198:32-42.

[22] Itakura H. Lipoprotein and lipoprotein subfraction. *Nippon Rinsho.* 2004 Dec;62 Suppl 12:38-41

[23] Moses RG, Calvert D, Storlien LH. Evaluation of the Accutrend GCT with respect to triglyceride monitoring. *Diabetes Care.* 1996 Nov;19(11):1305-6.

[24] Mahtabjafari M, Masih M, Emerson AE. The value of pharmacist involvement in a point-of-care service, walk-in lipid screening program. *Pharmacotherapy.* 2001 Nov;21(11):1403-6.

[25] Olson KL, Tsuyuki RT. Patients' achievement of cholesterol targets: a cross-sectional evaluation. *Am. J. Prev. Med.* 2003 Nov;25(4):339-42.

[26] Fodinger M, Sunder-Plassmann G, Wagner OF. Trends in molecular diagnosis. *Wien Klin Wochenschr.* 1999;111:315-9.

[27] Jain KK. Applications of biochips: from diagnostics to personalized medicine. *Curr. Opin. Drug Discov. Devel.* 2004;7:285-9.

[28] Robinson BH, Seeman NC. The design of a biochip: a self-assembling molecular scale memory device. *Protein Eng.* 1987;1:295-300.

[29] Odde DJ, Renn MJ. Laser-guided direct writing for applications in biotechnology. *Trends Biotechnol.* 1999;17:385-9.

[30] Kim DS, Lee SH, Ahn CH, Lee JY, Kwon TH. Disposable integrated microfluidic biochip for blood typing by plastic microinjection moulding. *Lab. Chip.* 2006;6:794-802.

[31] Soper SA, Hashimoto M, Situma C, Murphy MC, McCarley RL, Cheng YW, Barany F. Fabrication of DNA microarrays onto polymer substrates using UV modification protocols with integration into microfluidic platforms for the sensing of low-abundant DNA point mutations. *Methods.* 2005;37:103-13.

[32] Besselink GA, Vulto P, Lammertink RG, Schlautmann S, van den Berg A, Olthuis W, Engbers GH, Schasfoort RB. Electroosmotic guiding of sample flows in a laminar flow chamber. *Electrophoresis.* 2004;25:3705-11.

[33] Soo Ko J, Yoon HC, Yang H, Pyo HB, Hyo Chung K, Jin Kim S, Tae Kim Y. A polymer-based microfluidic device for immunosensing biochips. *Lab. Chip.* 2003;3:106-13.

[34] Ionescu RE, Marks RS, Gheber LA. Manufacturing of nanochannels with controlled dimensions using protease nanolithography. *Nano Lett.* 2005;5:821-7.

[35] Albers J, Grunwald T, Nebling E, Piechotta G, Hintsche R. Electrical biochip technology--a tool for microarrays and continuous monitoring. *Anal. Bioanal. Chem.* 2003;377:521-7.

[36] Cui HF, Ye JS, Chen Y, Chong SC, Sheu FS. Microelectrode array biochip: tool for in vitro drug screening based on the detection of a drug effect on dopamine release from PC12 cells. *Anal. Chem.* 2006;78:6347-55.

[37] Takahashi Y, Nagata T, Nakayama T, Ishii Y, Ishikawa K, Asai S. GeneChip system from a bioinformatical point of view. Nippon Yakurigaku Zasshi. 2002;120:73-84.

[38] Zhu T. Global analysis of gene expression using GeneChip microarrays. *Curr. Opin. Plant. Biol.* 2003;6:418-25.

[39] Ragoussis J, Elvidge G. Affymetrix GeneChip system: moving from research to the clinic. *Expert Rev. Mol. Diagn.* 2006;6:145-52.

[40] Jane KK. Applications of AmpliChip CYP450. *Mol. Diagn.* 2005;9:119-27.

[41] de Leon J, Susce MT, Murray-Carmichael E. The AmpliChip CYP450 genotyping test: Integrating a new clinical tool. Mol. *Diagn. Ther.* 2006;10:135-51.

[42] de Leon J. AmpliChip CYP450 test: personalized medicine has arrived in psychiatry. *Expert Rev. Mol. Diagn.* 2006;6:277-86.

[43] Dupuy AM, Lehmann S, Cristol JP. Protein biochip systems for the clinical laboratory. *Clin. Chem. Lab. Med.* 2005;43:1291-302.

[44] Molloy RM, Mc Connell RI, Lamont JV, FitzGerald SP. Automation of biochip array technology for quality results. *Clin. Chem. Lab. Med.* 2005;43:1303-13.

[45] Haglmuller J, Rauter H, Bauer G, Pittner F, Schalkhammer T. Resonant nano-cluster devices. *IEE Proc Nanobiotechnol.* 2005;152:53-63

[46] Sokuler M, Gheber LA. Resonant nano-cluster devices. *IEE Proc. Nanobiotechnol.* 2005;152:53-63.

[47] Degenhart GH, Dordi B, Schonherr H, Vancso GJ. Micro- and nanofabrication of robust reactive arrays based on the covalent coupling of dendrimers to activated monolayers. *Langmuir.* 2004;20:6216-24.

[48] Communication systems in healthcare. *Clin. Biochem. Rev.* 2006 May;27(2):89-98.

[49] Weiss RL. Effectively managing your reference laboratory relationship. *Clin. Leadersh. Manag. Rev.* 2003 Nov-Dec;17(6):325-7.

[50] Ash KO. Impact of cost cutting on laboratories: new business strategies for laboratories. *Clin. Chem.* 1996 May;42(5):822-6.

[51] Riley PA. Commercialization of health services: implications for the laboratories. *Malays. J. Pathol.* 1996 Jun;18(1):21-5.

[52] Wah DT, Wises PK, Butch AW. Analytic performance of the iQ200 automated urine microscopy analyzer and comparison with manual counts using Fuchs-Rosenthal cell chambers. *Am. J. Clin. Pathol.* 2005 Feb;123(2):290-6.

[53] Lamchiagdhase P, Preechaborisutkul K, Lomsomboon P, Srisuchart P, Tantiniti P, Khan-u-Ra N, Preechaborisutkul B. Urine sediment examination: a comparison between the manual method and the iQ200 automated urine microscopy analyzer. *Clin. Chim. Acta.* 2005 Aug;358(1-2):167-74.

[54] Hofer TP, Asch SM, Hayward RA, Rubenstein LV, Hogan MM, Adams J, Kerr EA. Profiling quality of care: Is there a role for peer review? *BMC Health Serv. Res.* 2004 May 19;4(1):9.

[55] Shimetani N. Communication of useful information from laboratory physicians to clinical physicians. *Rinsho Byori.* 2003 Apr;51(4):336-40.

[56] Charuruks N, Lekngarm P. Pilot project of blood gas and electrolyte analyzer network at King Chulalongkorn Memorial Hospital. *Chula Med. J.* 2006; 50: 831.

Chapter IX

Laboratory Information System: A Problematic Case Studies

Abstract

Laboratory information system (LIS) is widely used in laboratory medicine at present. Based on the concepts of network and database, the system is very useful in every phase of laboratory cycle. Although many advantages of the system are accepted, limitation is still detected and usually be problematic Here, the author reports two problematic cases due to computer of LIS.

Keywords: Laboratory information system, computer.

Introduction

It can be said that the most important aim of all medical laboratories is the generation and distribution of test results. However, test requests usually bring medical personnel a need for workload distribution and analysis. The medical laboratory requires a good system for consolidation, validation and reporting. It should also be noted that the turnaround time of these mentioned activities is a measure of the quality of laboratory service to its customers and this can directly affect the satisfaction of them. Presently, computer is used in many aspects including medicine. The application of computer technology in medical laboratory helps the laboratory process a lot. In the present technological era, the deployment of information technology and the management of information are rapidly growth areas in the health care service including medical laboratory service. There is no question that the laboratory information system (LIS) plays a pivotal role in controlling the performance of medical laboratory's activities [1 – 4]. LIS helps the laboratory reduce turnaround times by optimizing both the quality and efficiency of information supply and results production as well as distribution [1 – 4]. The system provides comprehensive information and organisational management capabilities, giving the laboratory complete control over all operations, from the creation of the initial request to the production and reporting out the final test results to the users, physician at medical ward [1 – 4]. To provide the most useful information, its role, structure, function and usefulness must be well understood by the laboratory personnel [1 – 4]. Recently laboratory information system (LIS) became a new widely used tool for laboratory management [1 – 4]. However some times, the limitation of the LIS can be seen. Here, the author discusses two problematic cases in laboratory medicine due to the problem of computation LIS system

Case Studies

Case 1

A computer was sent to the urinalysis unit of the central laboratory of the hospital. A patient complicated for an abnormal laboratory report. He notified that he got a report of hematuria with an additional comment that this should be due to menstruation. He felt upset and worried about this report, therefore, he sent a complain to the laboratory.

Case 2

An episode of computer network problem is noted in the laboratory. The result of laboratory Investigation of a patient cannot be validated and jammed in the system. Root cause analysis revealed that there are two confirmations on request of this patient. However, the laboratory workers cannot delete the repetitive from the system.

Discussion

LIS is widely used in medical laboratory at present (Figure 1). The full system cover request management at wand laboratory analysis and report management at laboratory and result report to the patient and physician management. The use of computer system and computerized devices in medical laboratory is growing and becoming general [5]. Process once be performed by people are increasingly being transferred to machines, which implied the concept of automation. Not only for the easier and less expensive laboratory management, but also for the better effectiveness of laboratory services [5]. LIS can be developed by the computational expert panels of each medical institution or bought as a package from vendors [5]. Based on the concepts of network and database, the system is very useful in every phase of laboratory cycle [6]. Although many advantages of the system are accepted, limitation is still detected and usually be problematic [6]. Although the system is automatic it still required humanistic control such as keying of the information. Safely management and quality control of LIS system is therefore necessary. Monitoring of error is still necessary.

Data input → data processing → Data output

Figure 1. General process of LIS.

For the first case, the problem is not the error of the system. The process is complete however the problem is due to the parameter setting of the result report section. Careful setting of the comment adding to the laboratory result is necessary.

Considering the second case, this is the real existence of network LIS problem. The jamming of the data from the repetitive input can be seen. Careful setting of allowance for keying of the data as will be deletion of the data is necessary. In addition, classification of degree of secret accompanied with specific password for corresponding worker is required [7].

Table 1. Safety management and quality control of LIS

	Network		
	input	process	output
Prevention of fire, stealing, breaking	Prevention of repetitive request, illegal request, in correct request	Prevention of incorrect analysis and report illegal analysis	Prevention of data hacking, secret of patient

Conclusion

LIS is widely used in laboratory medicine at present. Problems of LIS can be seen. An example of two problematic cases due to computer of LIS are hereby reported.

References

[1]   Gombas P, Skepper JN, Krenacs T, Molnar B, Hegyi L. Past, present and future of digital pathology. *Orv. Hetil.* 2004 Feb 22;145(8):433-43.

[2]   Buffone GJ, Moreau DR. Laboratory computing--process and information management supporting high-quality, cost-effective healthcare. *Clin. Chem.* 1995 Sep;41(9):1338-44.

[3]   Kanno T.From handmade to laboratory system. *Rinsho Byori.* 1997 Mar;45(3):203-5.

[4]   Charuruks N. Laboratory information systemIts role and importance in this era. *Chula Med. J.* 2000; 44 (4) : 229-242

[5]   Charuruks N. Laboratory information systemLIS in Thailand. *Chula Med. J.* 2000; 44 (5) : 319-337

[6]   Wiwanitkit V. Laboratory information management system : an application of computer information technology. *Chula Med. J.* 2000; 44 (11) : 887-891.

[7]   Fink R. Safety assessment of data management in a clinical laboratory. *Comput. Methods Programs Biomed.* 1994 Jul;44(1):37-43.

Chapter X

Ethics in Laboratory Medicine: Some Aspects

Abstract

Ethics is a basic principle in medicine. It must also be applied for laboratory medicine. In this article, the author will briefly detail and discuss on important aspects of ethics in laboratory medicine.

Keywords: ethics, laboratory medicine.

High Risk Laboratory Report: Problem and Prevention [1]

Presently, doctor generally makes use of laboratory report as a tool for decision making for patient care. There is an important kind of report that can be accepted as high risk laboratory report [2]. This group of reports is complicated reports that can lead to many psychological and social impacts on the patient. Examples are these problematic reports on oncological disorder, reports on human immunodeficiency virus test and reports on cytogenetic test [3 – 4]. Usually, there is a big problem regarding to the high risk laboratory report which can appear as the heading in the newspapers. In this article, the author hereby describes important aspects of high risk laboratory report.

1. Problem

Most problems are owing to violation of patient's right and improper broadcasting of patient's secret. The violation of patient's right generally starts from request without asking permission from the patient and giving no data for the patients [5]. The problem of exposed patient's secret is often owing to lade of appropriate secret keeping system. Often the patient's secret is exposed to the third party or broadcasted via the hospital computer network

[6]. In addition, there are also other additional problems owing to the error of laboratory personnel.

2. Prevention

As a preventive mean, informed consent must be strictly followed. The and post counseling should be applied and this activity can significantly reduce the problem. Setting of proper secret keeping that covers both document and online report is needed. Total quality management for the whole process is necessary.

Right Not to Know the Investigation Result [7]

Patient's right is the basic classical principle in medical ethics. Patient's right relating to laboratory result is the basic knowledge for all doctors. Basically, the two important rights [8 – 11] include

1. Right to get clinical data before investigation and to know the result
2. Right to have the test result kept secret.

In addition to these two major rights, there is another additional investigating right, right not to know the result.

This specific right of the patient is not well known. Routinely, the main of laboratory investigation is to derove result to help diagnosis or treatment. The investigation is the specific secret of the patient. The physician in change can know owing to the patient's trust. However, the patient who is the owner of the clinical information and secret has a complete right to know the laboratory test result [12]. In some cases, the patient might not desire to know his/her own laboratory test result but let the physician to investigate. This is the origin of the concept of "right not to know the investigation result." This right can be seen in laboratory test that has high social and psychological impacts such as anti HIV test [13 – 14]. Therefore, it is required that general practitioner should not only give information and ask for information prior to investigation but also if the patient would like to know the investigation result or not. In case that the patient says no, the practitioner in change has to keep the laboratory result as a special secret.

When a Medical Student Refuse to Get Procedure Training

The author got an interesting consultation from a faculty staff on a problematic case of a medical student that refuse to got medical procedure practice from her friend in a laboratory procedure practice class. The main questions are : 1. As a teaching staff, how to manage this case? and 2. Does a teaching staff have a right to force this student to accept the procedure practice by her friend? This situation is a human right, more than a simple patient's right. Basically, a patient can refuse to get any medical procedures of it is not in the actual emergency case. In this case, it seems to be medical student's right the topics to be discussed include:

1. Any medical procedures practice, for diagnostic or therapeutic purposes can be considered as a course of injury to the patient. The physician gets a good status to do necessary human to the patient aiming at caring and it is accept by legal and social concerns. However, the patient still has the right to reject [15]. In this case, the problematic person is a medical student, not a patient. Claiming for diagnostic purpose cannot be accepted. Medical procedure practice training among medical students is not the same as medical procedure practice training on the real patients under well supervised by teaching medical staff [16 - 17]. The case of procedure training among medical students seems to be an agreement to do human between medical students.

2. The following problem that the medical student rejects to get a training form his/her friend is to be concerned. Forcing can bring the complaint that it is an illegal practice. In addition, other medical students or teaching staff have no right to force.

3. Solving this problem should start from the medical student selection process [18 - 19]. Screening for the candidates with good attitude towards medical education system, acceptance to the fact that real medical practice on him/her self before on the real patient is needed, is required. In this case, the solution should be allowance the student's intension. The effective channel to solve the problem should be the action of the student's friends that also refuse this student's training on themselves and this can lead can lead to the final solution by the principle of group learning.

Violation on Medical Personnel in Laboratory: A Case Study

1. Introduction

In the present day, patient right is the heart of quality system in medicine [20 - 21]. This can also be applied for medical laboratory management. As to physicians, most criminal cases primary care physicians are usually due to failing to examine the patient or for diagnostic error [22]. On the other hand, the new concept on the medical personnel right is of interest in the present day [22]. At least medical personnel should have basic human right dining their daily practice in medical care. In this article, the author reports case studies on violation on medical personnel in a laboratory

2. Case Studies

Case 1: The in-charge nurse of the laboratory consulted the in-charge physician for the problem in working in venipuncture notification from a patient. Fro the history this patient visited to the veripuncture clinic on the morning send tried to violate the quota system of the clinic. The in-charge nurse tried to stop this patient and explained him to get in the quota system. But this patient felt very angry and start to give the bad words to the nurse. The in-charge nurse reported this incidence and also left a question if she was legally violated by those rude words.

Case 2 : The phlebotomist reported a case on the possible sexual violation on herself by a patient. She reported that she noted that this patient tried to focus his sight on her breasts and tried to forward his arm on the venipuncture desk and touched the phlebotomist breasts. The phlebotomist tried to explain and asked the patient to pulled his arm backward

3. Discussion

Medical personnel right is the new concept [23]. Similar to any occupation the violation on the medical personnel can be expected [24]. A challenge of particular significance is that encountered when a client develops an attraction to a medical worker. Education traditionally has not equipped medical personnel with the theoretical knowledge or experience to address this phenomenon in clinical practice [25]. In this article, two cases on violation on laboratory workers are reported. Indeed the blood collection unit of the laboratory is the port where the medical workers have direct interrelation with the patients. The first case is the case of the rude verbal complaint by a patient. Indeed, the verbal type of violation is common in medical service [26]. Similar to any service complaint is an important quality index in medical service and there is a need to set the system to get the patient's complaint. This patient gave a rude complaint and it is classified as a major risk to be solved [26]. The complaints of the patients on the mass media are totally unwanted episodes. Generally it is recommended that the medical worker should not directly response in a rude manner to the patient. Of interest, the question if rude complaint to somebody among others can be classified as illegal action and the practitioner has his/her right to sue the violator. However, this action should be the last choice.

Concerning the second case the problem of sexual violation on medical worker should be addressed. Since most of the nurses who directly contact with the patients in ward or clinic are female, the violation by verbal communication or direct touch can be expected. Actionable sexual harassment is defined as a violation of Title VII of the Civil Rights Act [27]. Primary management by asking the patient to stop the unwanted behavior as presented in thy second case is recommended. However if this attempt is not successful, call for help from the others as will as notification of hospital guard to charge the violator are suggested. Finally, specific protocol to respond with violation should be set. Medical personnel right should also be broad cast.

Conflict of Interest

1. What Is Conflict of Interest?

Conflict of interest is a widely used term in daily conversation. This can be due to two important etiologies:

1. Role conflict: This conflict is due to the fact that the practitioner has two roles in the same scenario. For example, the practitioner has the role as teacher for teaching medical student and has another role as physician to treat the patients.
2. Status conflict: This conflict is due to the fact that the practitioner has two statuses in the same scenario. For example, a practitioner is a member of board that makes decision in buying something and he/she is also the board of the commercial company which proposes for selling thing.

Conflict of interest will be not problematic if that conflict does not bring usefulness to the practitioner or other ones in a specific way. The problem of conflict of interest needs carefully managed and can be a problem in medical ethics.

2. Conflict of Interest in Medicine [28 - 29]

In medicine, conflict of interest might be not well known but it is very important because the medical commerce at present is a very big commerce. Millions dollars circulate in medical economy. There are only a few suppliers and some lack for specific medical knowledge. The chance for disguised information or falsified information giving to the client can be expected [30].

In medical science, conflict of interest is a basic thing to be declared in medical article. This is known as "Declare of conflict of interest". This is the international code and the conflict of interest that one/ones tend to support the interest of medical company needs to be clarified. Giving information on conflict of interest helps the readers to decide to believe on the given information in the article [31]. Conflict of interest is not problem but the problem is due to disguising [32].

3. Conflict of Interest in Laboratory Medicine

Similar to other branches of medicine, conflict of interest can be seen in laboratory medicine. This can be in these following forms.

3.1. Conflict of Interest in Scientific Aspect [33]

Similar to other kinds of medicine, medical information in laboratory medicine is needed for the practitioner. There are many kinds of information such as the information of the property of the reagent or analyzer. The reports on the reagent or analyzer must be clarified for the conflict of interest. The reader must carefully concern on the information on such reports sponsored by medical company

3.2. Conflict of Interest in Practice [34 - 35]

Often, the practitioner in medical laboratory get gift from representative of medical company. This can bring the bias in selection of reagent or analyzer. The receipt of gift must be well legally controlled.

References

[1]   Wiwanitkit V. High risk laboratory report: problem and prevention. *J. Med. Assoc. Thai.* 2008; 91 (7): 1146-7.

[2]   Nyrhinen T, Leino-Kilpi H. Ethics in the laboratory examination of patients. *J. Med. Ethics.* 2000;26:54-60.

[3]   Raymond D. AIDS, HIV testing, and the ethics of informed consent. *Ethics. Med.* 1987;3:9-15, 20.

[4]   Harris M, Winship I, Spriggs M. Controversies and ethical issues in cancer-genetics clinics. *Lancet Oncol.* 2005;6:301-10

[5]   Mallardi V. The origin of informed consent. *Acta Otorhinolaryngol. Ital.* 2005;25:312-27.

[6]   Yakata M, Okada M. Laboratory data management and medical ethics. *nsho Byori.* 1990; Spec No 83:5-10.

[7]   Wiwanitkit V. Right not to know the investigation result. *J. Med. Assoc. Thai.* 2008; 91 (4): 595 - 6.

[8]   Gupta S, Bhardwaj DN, Dogra TD. Confidential communications in medical care. *Indian J. Med. Sci.* 1999;53:429-33.

[9]   Mallardi V. The origin of informed consent. *Acta Otorhinolaryngol. Ital.* 2005;25:312-27.

[10]   Hull RT. Informed consent: patient's right or patient's duty? *J. Med. Philos.* 1985;10:183-97.

[11]   Trudeau ME. Informed consent: the patient's right to decide. *J. Psychosoc. Nurs Ment. Health Serv.* 1993;31:9-12.

[12]   Giesen D. The patient's right to know--a comparative law perspective. *Med. Law.* 1993;12:553-65.

[13]   Gray RH, Sewankambo NK, Wawer MJ, Serwadda D, Kiwanuka N, Lutalo T. Disclosure of HIV status on informed consent forms presents an ethical dilemma for protection of human subjects. *J. Acquir. Immune Defic. Syndr.* 2006;41:246-8.

[14]   Dhai A, Noble R. Ethical issues in HIV. *Best Pract. Res. Clin. Obstet. Gynaecol.* 2005 Apr;19(2):255-67.

[15]   Jones JW, McCullough LB, Richman BW. Informed consent: it's not just signing a form. *Thorac. Surg. Clin.* 2005;15:451-60.

[16]   Shooner C. The ethics of learning from patients. *CMAJ.* 1997;156:535-8.

[17]   Wiwanitkit V. What's missing from the class? *J. Med. Assoc. Thai.* 2000;83:577-8.

[18] Smith JK, Weaver DB. Capturing medical students' idealism. *Ann. Fam. Med.* 2006 Sep-Oct;4 Suppl 1:S32.

[19] Tilburt J, Geller G. Viewpoint: the importance of worldviews for medical education. *Acad. Med.* 2007 Aug;82(8):819-22.

[20] Weil V, Arzbaecher R. Ethics and relationships in laboratories and research communities. *Prof. Ethics.* 1995;4:83-125.

[21] Ittithumwinit C, Lertpadungkulchai S, Veerasoontorn V. Patients' legal rights. *J. Health Sci.* 1997; 6: 396-402.

[22] Varga T, Szabo A, Dosa A, Bartha F. Criminal liability of physicians and other health care professionals in Hungary (review of case law between 1996-2000). *Med. Law.* 2006;25:593-9.

[23] Thai Medical Council. Physician rights. Declaration No. 46/2006 30 November 2006.

[24] Randall T. Health professionals persecuted in violation of their human rights: a partial list of cases. *JAMA.* 1993;270:554.

[25] Schafer P. When a client develops an attraction: successful resolution versus boundary violation. *J. Psychiatr. Ment. Health Nurs.* 1997;4:203-11.

[26] Wiwanitkit V. Survey of satisfaction and complaint of customers of laboratory service, King Chulalongkorn Memorial Hospital. *Songkhanagarind Med. J.* 2002 ; 20: 85-89.

[27] Robinson RK, Franklin GM, Tinney CH, Crow SM, Hartman SJ. Sexual harassment in the workplace: guidelines for educating healthcare managers. *J. Health Hum. Serv. Adm.* 2005;27:501-30.

[28] Tonelli MR. Conflict of interest in clinical practice. *Chest.* 2007 Aug; 132(2): 664 – 70.

[29] Rabinovitch A. Financial conflict of interest? *Arch. Pathol. Lab. Med.* 2004 Dec; 128(12): 1328

[30] Rhodes R, Strain JJ. Whistleblowing in academic medicine. *J. Med. Ethics.* 2004 Feb; 30(1): 35 – 9

[31] Daicos GK. Ethical dilemmas encountered during clinical drug trials. *Hum. Health Care.* 2001 Jul – dec; 1(2): E9

[32] Baim DS, Donovan A, J Smith J, Briefs N, Geoffrion R, Feigal D, Kaplan AV. Medical device development: managing conflicts of interest encountered by physicians. *Catheter Cardiovasc. Interv.* 2007 Apr 1; 69(5): 655 – 64.

[33] Geller G, Bernhardt BA, Gardner M, Rodgers J, Holtzman NA. Scientist's and science writers' experiences reporting genetic discoveries: toward an ethic of trust in science journalism. *Genet. Med.* 2005 Mar; 7(3): 198 – 205.

[34] Harty-Golder B. Laboratory limits. *J. Fla. Med. Assoc.* 1996 Mar; 83(3): 201 – 3.

[35] Dechene JC, Howton T. Lab discounts. How deep is too deep? *CAP Today.* 2000 Mar; 14 (3): 42 – 4, 48

Chapter XI

Urine Strip Examination, Interesting Problematic Cases in Laboratory Medicine

Abstract

Urine examination is a basic performed in every medical laboratory. For urine strip examination, limitations of the test are mentioned. Here, the anther presented four interesting problematic case of urine strip examination in laboratory medicine.

Keywords: urine strip, problem.

Introduction

Urine examination is a basic performed in every medical laboratory. For the basic urinalysis, macroscopic, microscopic and urine chemistry tests are performed [1 - 2]. For urine chemistry test, urine strip examination is applied [1 – 2].

For urine strip examination, limitations of the test are mentioned [3 - 4]. Here, the anther presented two interesting problematic case of urine strip examination in laboratory medicine.

Case Reports

Case 1

A discussion of the urine strip examination procedure is made in quality round. The discussed topic is about the result of urine strip from reading of the strip with and without preprocessed by swabbing at the mine collection bottle. It can be seen that the results form both method are totally different.

Case 2

A physician in change was notified for the abnormal mine strips examination result from the medical technology in change. The problem occurred in the cases of urine sample of diabetic portents with pyuria. The medical technologist noted that if the urine strip was processed and read directly, the result would be negative for sugar and ketone. However if the mine sample was centrifuged to sediment the white bleed cell, the result would be positive for either sugar of ketone.

Case 3 [5]

A female patient consulted to the in charge physician about urine pregnancy strip test. She said that she bought a test kit from a pharmacist at a drug store and performed the test in the morning on the following day and got negative result, presenting with one line in the strip. However, she left the strip on the desk and she accidentally found that the left test strip became positive, presenting with two lines in the urine strip on that night. She then felt very frustrated and thought that there might be somethings erro. Therefore, she went to the physician in charge for recieving medical consultation. She requested for the urine pregnancy test examination to confirm and the result was negative, presenting with one line in the strip. The final confirmation test was negative for urine pregnancy result.

Case 4 [5]

A diabetic patient presented his diabetic self record to the in charge physician and the daily urine glucose strip test results were negative in all records. However, the capillary glucometer results revealed high glucose levels, which is totally discordant. The discordant results were investigated for the possible root cause and the physician finally found that the patient falsely dip the strip and abruptly read the result in a shorter period of time than what is suggested in the leaflet or that commercial test kit.

Discussion

Urine strip examination is a basic medical laboratory analysis [6 - 7]. In this article, the first case can be a good example on pre-analytical error. In the correct procedure, swabbing is needed. If there is no swabbing the lift large droplet of urine covering the strip can bring incorrect aberrant results. This error seems simple but usually be overlooked and can result in aberrant laboratory results [3 - 4]. Quality control in the urinalysis procedure is necessary [3].

In the second case, more complicated finding is reported. The interference of white blood cell bringing opacity of the sample can distorted the urine strip examination result [3, 8]. This can also be found even in case of automated urine strip examination. The big problem in this case is that pyuria is a common complication in diabetic portent and if the interference occurs the misdiagnosis of glycosuria and ketonuria which are also the two common findings in complicated diabetic patients.

The third case is an example of false positive due to prolonged result reading [5]. Routinely, urine test strip for pregnancy test should be read at 3 minutes after dipping [9]. The case of false positive is very interesting in laboratory medicine [5]. If the urine strip is left on the desk, exposing to the environmental atmosphere, there might be contamination for other chemicals from the environment and can lead to the aberration of results [5]. Also, it might be probable that it was the result of redox reaction, therefore, further studies on this situation are recommended [5]. More than 355,000 urine pregnancy tests were sold to the public, pharmacies, laboratories, and physicians [10]. Suhr said that public education by pharmacies could help confirm with at least a 95 % reliability rate, especially if the second test was performed within 1-2 weeks after the 1st test was taken [10]. These tests are reliable not only when they are taken at the right time, but also through consultation with a physician to assure certainty about actual pregnancy [10].

In the last (fourth) case, the patient immediately read the result without waiting period for the reaction to complete [5]. Therefore, the false negative can be seen. However, there are other probable causes of false negative urine glucose test [5]. Several drugs or drug classes have been well documented to clinically interfere with these tests [5]. The documented interfering drugs include ascorbic acid, beta-lactam antibiotics (e.g., cephalosporins and penicillins), levodopa, and salicylates [11]. Several drugs or their metabolites that are strong reducing substances produce false-negative results by the glucose oxidase method [11]. However, there was no evidence of drug usage in this patient, therefore, it is needed to assure that the patients understood the correct procedure of this monitoring test before letting them perform on their own tests [5].

Reference

[1]   Jinde K, Endoh M, Sakai H. Usefulness of urine analysis in clinical nephrology. *Rinsho Byori.* 2003 Mar;51(3):214-8.

[2]   Ito K. Recent advances on routine urinalysis. *Rinsho Byori.* 2000 Sep;48(9):823-8.

[3]   Orita Y, Ito K, Igarashi S, Koba T, Shimada I, Imai N. Report on the accuracy and the reliability of dip sticks. *Rinsho Byori.* 1996 Nov;44(11):1100-11.

[4]   Ekawong P, Wiwanitkit V. Limitations of urine strip method. *Chula Med. J.* 2005 49 (8): 437-443.

[5]   Wiwanitkit V. Problematic cases of urine strip examination: a time effect. *Chula Med. J.* 2008 Jan - Feb; 52(1): 59 - 62.

[6]   Wilson LA. Urinalysis. *Nurs. Stand.* 2005 May 11-17; 19(35): 51-4.

[7]   Larsson L, Ohman S. Quality assurance in urine analysis. *Qual. Assur. Health Care.* 1992 Jun;4(2):141-9.

[8]   Gillenwater JY. Detection of urinary leukocytes by chemstrip-l. *J. Urol.* 1981 Mar;125(3):383-4.

[9]   Chard T. Pregnancy tests: a review. *Hum. Reprod.* 1992 May; 7(5): 701-10 10.

[10] Suhr P. Pregnancy tests. Can we trust them? *Ugeskr Laeger.* 1983 Jul 4; 145(27): 2104-
      6.

[11] Rotblatt MD, Koda-Kimble MA. Review of drug interference with urine glucose tests.
      Diabetes Care. 1987 Jan-Feb;10(1):103-10.

Chapter XII

Light Microscopy:
Past, Property and Procedure

Abstract

Visualization is the aim in science. Due to the limitation of eye sight, there is a need to get scientific tool that helps scientist see or visualize small objects. The origin of microscope was in 1600. In this article, past history of microscope is presented. In addition, technical properties of microscope are discussed and the light microscope procedures, both examination procedure and cleaning and care of the microscope are mentioned.

Keywords: microscope.

Origin of Light Microscope:
Past History [1-7]

Visualization is the aim in science. Due to the limitation of eye sight, there is a need to get scientific tool that helps scientist see or visualize small objects. Those small objects, which cannot be visible in naked eyes, in medicine include bacteria, fungus and small parasites. The origin of microscope was in 1600. The light microscope opened the mind to the scientist of the hitherto unseen. Its acceptance was a seminal event in history that helped scientist to conceive the inconceivable and encouraged them to think the unthinkable. The first microscope was essentially simple magnifiers, although some achieved quite high powers. Much important work was done with simple microscopes well into the next centuries. In 17th century, flea glasses were small simple microscopes, made of ivory, bone, wood or tortoise shell, used to study insects impaled on a pin. Antoni van Leeuwenhoek's observations through his high power simple microscopes were communicated to the Royal Society in London and this is accepted as the father step of light microscope [8 – 10]. Leeuwenhoek's microscope is a glass bead lens sandwiched between two plates of brass and a studied specimen could be affixed to a spike and held close to the eye for examination [8 –

10]. In 19th century, more updated simple microscope was built by Robert Banks and this is the origin of English microscope. As a new finding in this era, Robert Brown observed the cell nucleus, cytoplasmic streaming (cyclosis), and reported a famous phenomenon now known as Brownian motion [11]. However, due to the great complexity and expense of English microscopes, the Society of Arts in 1854 had offered a prize for a practical and affordable student microscope and the winner was Robert Field. The pattern of Robert Field is the classical pattern at present. At present, there are several varieties of light microscope including bright-field microscope, dark-field microscope, phase-contrast microscope and fluorescence microscopes.

Important Technical Properties of Light Microscope [12 – 14]

The heart of light microscope is "lense". Lenses and the bending of light is the basic fundamental of microscopy: light is refracted or bent when passing from one medium to another. The refractive index, a measure of how greatly a substance slows the velocity of light, is for determination for direction and magnitude of bending of the two media forming the interface. Lenses are for focusing light rays at a specific place called the focal point and such distance between center of lens and focal point is called focal length. It should be noted that strength of lens related to focal length which can be demonstrated as long focal length means less magnification. Finally, total magnification is the product of the magnifications of the ocular lens and the objective lens. Considering magnification, an object can be focused generally no closer than 250 mm from the eye and this is considered to be the normal viewing distance for 1x magnification. Modern microscopes magnify both in the objective and the ocular and thus are called "compound microscopes". Bright field illumination does not bring differences in brightness between structural details while structural details can be derived via phase differences and by staining of components. In addition, the edge effects (diffraction, refraction, reflection) produce contrast and detail. At present, the microscope appearance is according to the pattern of Robert Field as previously mentioned. The standard of light microscope is originally conformed to the German DIN standard which requires the following; a) real image formed at a tube length of 160mm, b) the parfocal distance set to 45 mm and c) object to image distance set to 195 mm.

Table 1. Some properties of light microscope

Property	Definition
Resolution	Resolution means the ability of a lens to separate or distinguish small objects that are close together and the wavelength of light used is major factor in resolution.
Absorption	When light passes through an object the intensity will be decreased depending upon the color absorbed.
Refraction	Refraction means direction change of a ray of light passing from one transparent medium to another with different optical density.
Diffraction	When light rays bend around edges, new wavefronts will be generated at sharp edges.
Dispersion	Dispersion means separation of light into its constituent wavelengths when entering a transparent medium

Light Microscope Procedure [12 – 14]

A. Examination Procedure

Light microscope procedure is a basic technique that every physician must perform. It should be noted that most of specimens examined are "wet" specimens such as urine sediments or wet mount preparations and these specimens are usually placed on a glass slide with a cover slip in place. Cover slipped slides should only be viewed using the low or high power objective with subdued light but do not use the oil immersion objective to examine a wet specimen. Before placing the cover slipped slide on the stage, it is necessary to make sure the low power objective is securely in place. Use the course focus knob to bring the maximum distance between the objective and the stage. When adjusting the course focus knob, look at the stage and if adjusted too fast the stage could make contact with objective and break the lenses or the slide if it happens to be in place. Then set the light source on the lowest setting possible and set the condenser at lowest setting. It is also necessary to make sure diaphragms are all open. Then adjust the eyepieces, course and fine focus knobs and the light source as necessary as necessary. When examine, scan the specimen under low power first and when an area of interest is derived, switch to the high dry objective for a closer look. If necessary, the light source may be increased but wet specimens are best viewed in subdued light. As a rule, examine at least 10 fields, continually adjusting the fine focus. After already record the results, remove the slide and dispose of it immediately in a sharps disposal container and wipe the stage area if any liquid remains.

B. Cleaning and Care of the Microscope

Daily maintenance of microscope is needed; users have to clean the eyepieces, the objectives and the light source at the end of each day or as needed during use.

For cleaning of lenses, either lens paper, silk or muslin cloth is allowed (alcohol or cotton swabs are noted allowed). For other parts, soap solution can be used. Basically, the lenses or eyepieces may easily become dirty and it should be noted that eye make-up may build up on the lenses or glasses may scratch them. In addition, when the microscope is not in use, it is best to turn the light source off and place the dust cover over the instrument. Store the microscope under a protective cover and in a low humidity environment. For at least once per year the microscope should be serviced by a trained professional. In technical maintenance, it is needed to clean the scope inside and out. They will also be able to make any necessary repairs.

Table 2. Cleaning process for light microscope

Step	Details
Step 1: Cleaning the Eyepieces	Blow to remove dust before wiping lens then clean the eyepieces with a cotton swab moistened with lens cleaning solution. Finally, clean in a circular motion inside out
Step 2: Cleaning the Objectives	Moisten the lens paper with the cleaning solution and wipe gently the objective in circular motion from inside out. Finally, wipe with dry tissue or lens cleaning paper. It should be noted that objectives should never be removed from the nosepiece.
Step 3: Cleaning the Microscope Stage	Wipe the microscope stage using the cleaning solution on a soft cloth then thoroughly dry the stage.
Step 4: Cleaning the Microscope Body	Unplug the microscope from power source and moisten the cotton pad with a mild cleaning agent. Finally, wipe the microscope body to remove dust, dirt, and oil.
Step 5: Cleaning the Condenser and Auxiliary Lens	Unplug the microscope from power source then clean the condenser lens and auxiliary lens using lint-free cotton swabs moistened with lens cleaning solution. At last, wipe with dry swabs.

References

[1]   Oster M. Seeing and doing. [Review of: Wilson C. The Invisible world: early modern philosophy and the invention of the microscope. Princeton University Press, 1995. *Ann. Sci.* 1996 Nov;53(6):617-25.

[2]   Bradbury S. The history of the reflecting microscope. *Proc. R. Soc. Med.* 1969 Jul 7;62(7):673-4.

[3]   Holck P. From the history of microscope. *Tidsskr Nor Laegeforen.* 1988 Mar 20;108(8-9):654-9.

[4]   Fournier M. Dutchmen and the development of the microscope. *Ned. Tijdschr. Geneeskd.* 1991 Dec 21;135(51):2433-8.

[5]   Haselmann H. The microscope, tool and object of science. *Z. Wiss. Mikrosk.* 1966 Dec;67(4):244-56.

[6]   Amos B. Lessons from the history of light microscopy. *Nat. Cell. Biol.* 2000 Aug;2(8):E151-2.

[7]   Allen E, Turk JL. Microscopes in the Hunterian Museum. *Ann. R. Coll. Surg. Engl.* 1982 Nov;64(6):414-8.

[8]   Shklar G. Leeuwenhoek and Vermeer, an association of genius. *J. Hist. Dent.* 1998 Jul;46(2):53-7.

[9] Casida LE Jr. Leeuwenkoek's observation of bacteria. *Science.* 1976 Jun 25;192(4246):1348-9.

[10] Faludy A. Antoni van Leeuwenhoek and the simple microscope. *Orv. Hetil.* 1973 Oct 7;114(40):2429-31.

[11] Kubo R. Brownian Motion and Nonequilibrium Statistical Mechanics. *Science.* 1986 Jul 18;233(4761):330-334.

[12] Light Microscopy. Available online at http://www.ruf.rice.edu/~bioslabs/methods/ microscopy/microscopy.html

[13] Microscope. Available online at www1.stkc.go.th/stportalDocument/ stportal_ 1170654028.doc

[14] Microscope Terms. Available online at www.opticsplanet.net/microscope-terms.html

Chapter XIII

Concerns on Vacuum Tube

Abstract

Vacuum tube is the widely collection container for blood. There are several kinds of vacuum tube for usage. In this article, the author briefly mention for concerns on vacuum tube.

Keywords: vacuum, tube.

Preanalytical Concerns on the Monosed-SR for the New Microsed-SR Erythrocyte Sedimentation Rate Analyzer

1. Introduction

The erythrocyte sedimentation rate (ESR) is a time-honored blood test, which assesses the degree of erythrocyte aggregation by acute phase proteins such as fibrinogen and immunoglobulins [1]. The ESR still is a very valid test for the diagnosis of certain chronic diseases (polymyalgia, rheumatoid arthritis, multiple myeloma, septic arthritis and ostemyelitis) and the follow-up of certain chronic diseases (polymyalgia rheumatica, systemic lupus erythematodes, chronic infections, prostatic cancer, and Hodgkin's disease) [1]. It is simple, inexpensive test, but unfortunately it lacks sensitivity and specificity [2]. Clinicians need to be aware of appropriate uses, because any test is expensive when ordered often, and evaluation of false-positive results may incur substantial costs and place the patient at risk from additional procedures [2].

There are several new techniques developed for determination ESR at present [3]. Automated ESR analyzer have been launched for a few years and approved for the laboratory diagnostic property [4]. However, the real usage of the system must concern the whole process starting from the pre-analytical phase. Here, the authors reported the mechanical consideration and pre-analytical user response to the new Automated ESR analyzer.

2. Materials and Methods

Automated ESR Analyzer

A new erythrocyte sedimentation determination (ESR) method, MicroSed SR-system (ELECTA-LAB), was evaluated for its appropriateness in real usage in the clinical laboratory. This system is the new analyzer based on the piezoelectrical principle. The sedimentation rate can read within 30 minutes. The system has to be used with a specific vacuum tube, MonoSed –SR containing 0.13 % sodium citrate. This tube will be further considered in the evaluation of pre-analytical phase.

Consideration for the Pre-Analytical Phase

According to the theoretical principle, the consideration for the pre-analytical phase should be focus on both mechanical consideration as well as user response.

A. Mechanical consideration

The size (length and diameter) of specific vacuum tube for the MicroSed SR-system is compared to that of the standard vacuum tube as well as the needle holder.

B. User response

The response of the pre-analytical users or phlebotomists at the venipucture clinic of King Chulalongkorn Memorial Hospital after 1-month trial was survey by interviewing and summarized.

Ideal Tube for Usage

Ideally, the most appropriate tube must fit into the standard needle holder and match the fully extension span between thumb and index fingers of the user.

3. Results

A. Mechanical consideration

The size of MonoSed –SR, classical ESR vacuum tube and standard needle holder are shown in Table 1. Comparing to the standard needle holder, classical ESR vacuum tube has 0.5 cm smaller in diameter and 3 cm longer in length. Comparing to the standard needle holder, MonoSed –SR vacuum tube has 1.0 cm smaller in diameter and 5.0 cm longer in length.

B. User response

The trial of the MonoSed –SR was set at venipucture clinic of King Chulalongkorn Memorial Hospital. The response of the pre-analytical users or phlebotomists at the venipucture clinic of King Chulalongkorn Memorial Hospital after 1-month trial was summarized. All of the phlebotomist (8 phlebotomists) noted that this new vacuum tube is not comfortable for usage due to two reasons; a) the length of the tube exceeds the fully extension span between thumb and index fingers and b) the diameter of the tube is too small comparing to the standard needle holder. They said that these two reasons bring the delay in the venipuncture process for 1 – 2 minutes/ cases.

**Table 1. Mechanical consideration of MonoSed –SR, classical
ESR vacuum tube and standard needle holder**

equipment	Diameter (cm)	Length (cm)
MonoSed-SR	1.0	12.5
Classical ESR vacuum tube	1.5	8.5
Needle holder	2	5.5

* Average fully extension span between thumb and index fingers of general population is about 11 – 13 cm.

4. Discussion

Erythrocyte sedimentation rate is a non-specific parameter used for the differential diagnosis and follow-up the patients [1 – 3]. In the present day there are many methods to determine the erythrocyte sedimentation rate. All methods have the same principle - sedimentation principle [3]. The standard method is Classical Westergren method [3]. There are equipment developed in order to increase safety and reduce time required for the procedure [3].

In order to accept a new technique to the medical laboratory, a careful assessment is needed. The assessment process should focus not only analysis phase but also other phase of laboratory quality cycle. According to this work, the authors assess the pre-analytical phase acceptability of the new automated ESR analyzer. The author focused the interest on the new vacuum tube, MonoSed-SR. Here, it can be shown that the user perception to the new vacuum tube is not as good as the previous classical ESR tube. Indeed, Wiwanitkit recently indicated that the synchronized equipment and sharp needle are the two major requirements of the practitioners [5]. In this study, all phlebotomist experience the difficulty in using of the new tube due to the inappropriate length.

Mechanically, the MonoSed-SR is longer than the classical tube for 4 cm and it reach the fully extension span between thumb and index fingers of general population which can make the user difficulty in one hand management of the vacuum tube in the venipuncture process, especially for the step of tube insertion into the needle holder and mixing up of the blood specimen inside the tube. In addition, the tube length of MonoSed-SR is about 2 times greater than the length of the needle holder and the diameter of MonoSed-SR is about a half of that of the needle holder. These facts make the Mono-Sed-SR tube lies unstably within the holder and make the phlebotomist hard to fix. Mechanical fulcrum can be expected and this can lead the tube accidentally pushed outside the needle holder during venipucture before proper amount of specimen is derived and make that venipuncture fail. These situations are also notified by the phlebotomist in this study.

Comparison between Evacuated Blood Collection System and Classical System for Prothrombin Time Determination [6]

Introduction

Prothrombin time (PT) determination is widely used laboratory coagulation test in medical practice [7 – 9]. Routinely, the specimens for PT tests are venous blood specimens collected by venipuncture procedure. Presently, there are two major venipuncture methods: evacuated blood collection system and the syringe blood collection system. Concerning the evacuated blood collection system, blood automatically flow to react with anticoagulant in the vacuum tube whereas the syringe blood collection system based on a suction principle [10]. The advantage and limitation of the venipuncture techniques have been mentioned, however, there are still many factors affecting the specimen presentation. These include both equipment factors and user factors. Therefore, Wiwanitkit performed a study was determine the final result of the two techniques to compare how effective they are [6]. The study was generated as a retrospective descriptive study and Wiwanikit studied all available laboratory records in a one year period [6]. All specimens were considered using criteria of specimen rejection [10 – 11] (Table 2) to determine how proper the specimen presentations were [8]. The ratios of improper specimens in the evacuated blood collection system group and the syringe blood collection system group were 50/5124 and 102/3312 respectively showing that there was a significant difference of ratio between the two groups [6]. Wiwanitkit concluded that the evacuated blood collection system was appropriate for collection of blood specimen for PT test [6].

Table 2. Standard criteria of specimen rejection for PT test

causes	example
1. Improper in quantity	• Too much
	• Too little
2. Improper in quality	• Clot
	• Hemolysis

References

[1]   Reinhart WH. Erythrocyte sedimentation rate--more than an old fashion? *Ther. Umsch.* 2006 Jan;63(1):108-12.

[2]   Brigden M. The erythrocyte sedimentation rate. Still a helpful test when used judiciously. *Postgrad. Med.* 1998 May;103(5):257-62, 272-4.

[3]   Wiwanitkit V, Siritantikorn A. Methods to determine erythrocyte sedimentation rate in the present day. *Chula Med. J.* 2002 Jan; 46 (1): 87-102.

[4]   Wiwanitkit V, Chotekiatikul C, Tanwuttikool R. MicroSed SR-system: new method for determination of ESR--efficacy and expected value. *Clin. Appl. Thromb. Hemost.* 2003 Jul;9(3):247-50.

[5]   Wiwanitkit V. Evacuated blood collection system devices, what is the ideal in of opinion of the phlebotomists. *Yasothorn Med. J.* 1999 Jan - Apr; 2(1): 51-52.

[6]   Wiwanitkit V. Rejection of specimens for prothrombin time and relating pre-analytical factors in blood collection. *Blood Coagul. Fibrinolysis.* 2002 Jun;13(4):371-2.

[7]   Kim A, Musgrave A, Douglas A, Triplett DA. Quality assurance in the hemostasis laboratory. In: Musgrave A, Triplett DA, eds. Hematology Clinical and Laboratory Practice. 1st ed. St. Luois: Mosby , 1993: 1309 – 16.

[8]   Brown BA. Coagulation. In: Brown BA, eds. Hematology: Principles and Procedures. 6th. Pennsylvania: Lea & Fabiger, 1993 : 203 – 78.

[9]   George JN. Classification & clinical manifestation of disorders of hemostasis. In: Ernest B, Lichtmann MA, Coller BS, Kipps TJ, eds. Williams Hematology. 5th ed. New York: McGraw-Hill , 1995 : 1276 – 81.

[10] Kapit W, Macey RI, Meisami E. The physics of blood flow. In: Kapit W, macey RI, Meisami E, eds. The Physiology Coloring Book. New York: Harper Collins, 1987: 32.

[11] Young DS, Bermes EW. Specimen collection and processing : source of biological variation. In: Burtis CA, Ashwood ER, eds. Clinical Chemistry. 4th ed. Philadelphia: WB Saunders, 1996: 33 – 52.

Chapter XIV

Blood Glucose Determination in Diabetic Care: A Concern in Laboratory Medicine

Abstract

For diabetes care, laboratory investigation is a necessary part. Similar to general laboratory investigation, preanalytical error can be seen in the diabetic laboratory investigation. In this article, dangers of preanalytical error on diabetes care are discussed. Case study on preanalytical error for oral glucose tolerance test is also shown. The author recommends that collaboration between laboratory and ward medical personnel is necessary in coping with a diabetic patient presenting with aberrant laboratory result. Also, the author present the conceptual study on the effect of poor controlled diabetes defined by hemoglobin A1C test via hemoglobin Cranston model as annex to this article.

Keywords: glucose, diabetes, care.

Introduction

Following the development and successful implementation of high-quality analytical standards, analytical errors are no longer the main factor influencing the reliability and clinical utilization of laboratory diagnostics [1]. Errors occurring within the extra-analytical phases are still the prevailing source of concern [1]. In the recent studies, most errors occur in the preanalytical phase suggest the implementation of a more rigorous methodology for error detection and classification and the adoption of proper technologies for error reduction [2 - 3].

For diabetes care, laboratory investigation is a necessary part. Fasting plasma glucose is the standard test for diagnosis of diabetes mellitus [4 – 7]. Current research is aiming to define the blood glucose levels at which risks increase so that clinical management can be appropriately directed. The hallmark of untreated diabetes mellitus is a raised blood glucose concentration (hyperglycaemia), associated with many other disturbances of blood biochemistry, including acid-base and electrolyte imbalance [8]. This has confirmed a central role for clinical laboratories in both diagnosis of diabetes and lifelong monitoring of the diabetic patient [8]. Similar to general laboratory investigation, preanalytical error can be seen in the diabetic laboratory investigation. In this article, dangers of preanalytical error on diabetes care are discussed.

Dangers of Preanalytical Error on Diabetes Care

Preanalytical errors in diabetic laboratory investigation can be expected at any clinical laboratory (Table 1). A random error, which is usually due to human error, is the most common type of laboratory error [9]. Dangers of preanalytical error on diabetes care will be described.

A. Errors in Patient Preparation

Patient preparation is the first step of the laboratory process. Basically, fasting blood sample is required before fasting blood sugar determination. The information on fasting practice must be given to the diabetic patients [4 – 7]. However, the most common problem in patient preparation is poor compliance of the diabetic patient. The patients might disguise to the phlebotomist that they have good fasting practice. The abnormal high blood glucose result can lead to improper increase dosage of medication if the physician in charge does not concern. This can bring dangers from over dosage of diabetic drugs or insulin. In addition, the dangers of the fasting practice for the diabetic patient should also be concerned. The case of sudden due to severe hypoglycemia in a diabetic patient waiting for the blood test was also reported [10].

B. Errors in Patient Identification

Patient safety is influenced by the frequency and seriousness of errors that occur in the health care system [11]. Error in patient identification can be expected in the venipucture process and if it is occur, the aberrant laboratory result can be expected [12]. The improper diabetic care can be prescribed and unwanted complication can be expected. About 0.05 % of overall blood collections in the laboratory are reorted to have errors in patient identification [11, 13]. Technology, ranging from bar-coded specimen labels to radio frequency identification tags, can be incorporated into protective systems that have the potential to detect and correct human error and reduce the frequency with which patients and specimens are misidentified [13].

C. Errors due to Wrong Specimen Additives and Order of Collected Sample

Basically, fluoride is the recommended additives for blood glucose determination since fluoride can well preserve the glucose level. However, the wrong specimen additives might be used and this can affect the laboratory results. In a previous study in a medical faculty, only a few medical students could describe complete rational tube preparation correctly, therefore, the errors due to wrong specimen additives can be expected in medical practice [14]. In addition, for the oral glucose tolerance test (OGGT), which four blood samples are required, the phlebotomist might arrange the request forms and the collected blood samples with incorrect order.

D. Errors in Specimen Transportation

Similar to the patient identification in blood collection step, the error can also be seen in specimen transportation process. Teerakanchana et al studied of serum glucose in the same patients from 9 wards at different distance and found that the reduction of glucose level related to transportation system [15]. Indeed, the transportation of any specimens for blood glucose monitoring has to be performed within half an hour after blood collection [4 – 7].

E. Errors due to Biological Variability

Biological variability can be a source of abnormal laboratory result. The hyperemesis gravidarum is the main interference of the OGTT. OGTT can result in the abnormal low glucose level and can be an unacceptable aberrant laboratory result [16]. In additional to hyperemesis gravidarum, the inference from drug is another important factor contributing to aberrant OGTT result. Those drugs include thiazide diuretics, Beta-blockers, oral contraceptives and prednisone.

Table 1. Summarization of preanalytical error for diabetic laboratory investigation

Possible causes of error	Possible underlyings	Possible preventive action
1. no fasting	Poor patient compliance	Good information providing
2. incorrect patient identification	Malpractice of phlebotomist	Venipuncture guideline
3. improper specimen transportation	Malpractice of phlebotomist	Specimen transportation guideline

How to Control the Preanalytical Error in Diabetic Laboratory Investigation

Indeed, quality control is necessary for all laboratory tests. The control in pre-analytical, analytical and post-analytical period for diabetic laboratory investigations is recommended.

Most of the problematic case for diabetic test is in the pre-analytical phase. Whereas the preanalytical phase can be theoretically controlled completely in and by a laboratory, the problems usually exist. The application of laboratory quality systems such as ISO 15189 can be a tool to decrease the incidence of errors [17].

Prevention and corrective actions must be set. In addition, collaboration between laboratory and ward medical personnel is necessary in coping with a diabetic patient presenting with aberrant laboratory result.

A Case Study on Pre-Analytical Factor Interference for Oral Glucose Tolerance Test

Basically, glucose is the sugar that the body uses for energy. In glucose homeostasis, both hormone and enzyme play altogether to regulate the glucose metabolism. Considering the hormone, two hormones, insulin and glucagon from the pancreas play major role in glucose homeostasis. The abnormality in the glucose homeostasis results in several clinical disorders and the most well-known clinical disorder is diabetes mellitus. For diagnosis of fasting plasma glucose is the standard [18 – 21]. Current research is aiming to define the blood glucose levels at which risks increase so that clinical management can be appropriately directed. When available, the criteria required to justify population screening should be satisfied. However, there is another functional test, which can give more details on the glucose homeostasis in the patients [18 - 21]. The quoted test is the glucose tolerance test. The most common glucose tolerance test is the oral glucose tolerance test (OGTT). In clinical practice, OGTT is used in obstetrics. The OGTT is used to screen pregnant women for gestational diabetes between 24 and 28 weeks of pregnancy [18].

A case of aberrant OGTT laboratory result was reported to the physician in-charge at the clinical laboratory. In this case, the patients have the results of 75 gm OGTT as fasting = 312 mg/dL, 1 hour = 85 mg/dL and 2 hours = 216 mg/dL. The obstetrician suspect for some possible errors in the laboratory analysis, therefore, he sent an incident report on this episode to the laboratory.

From the root cause analysis, all of the laboratory analysis was performed with in a clinical chemistry laboratory using a standard clinical chemistry analyzer. Further history taking, the patients has well preparation for the OGTT and had no external identified source of interference. From further review, the error can be seen at the venipuncture clinic of the hospital. The phlebotomist arranged the request forms and the collected blood samples with incorrect order.

Indeed, quality control is necessary for all laboratory test. The control in pre-analytical, analytical and post-analytical period for glucose tolerance test is recommended. Most of the problematic case for glucose tolerance test is in the pre-analytical phase. The hyperemesis gravidarum is the main interference of the OGTT. OGTT can result in the abnormal low glucose level and can be an unacceptable aberrant laboratory result [22]. In additional to hyperemesis gravidarum, the inference from drug is another important factor contributing to aberrant OGTT result. Those drugs include thiazide diuretics, Beta-blockers, oral contraceptives and prednisone.

In this case, the preanalytical error can be seen. It is a good example of preanalytical error in OGTT, which can be expected at any clinical laboratory. Indeed, the preanalytical error is the most common type of error in laboratory process [23]. In this case, all of the analyses in the laboratory were correct but the defect is at the venipuncture clinic. This is a random error, which is usually due to human error [24].

In this case, the phlebotomist had several malpractices. First, the patient identification was severely missed. Second, the transportation of the specimen was also incorrect. Indeed, the transportation of any specimens for blood glucose monitoring has to be performed within half an hour after blood collection. This case can be a good case study for quality control in OGTT test in any laboratory. The summarization of possible preanalytical error for OGTT is presented in Table 1.

A Conceptual Study on the Effect of Poor Controlled Diabetes Defined by Hemoglobin A1C Test via Hemoglobin Cranston Model

1. Introduction

Diabetes mellitus is a frequent disorder affecting subjects of all ages [25]. This disease is associated with significant morbidity and fatality and increasing medical costs, and its prevalence is increasing to concerned epidemic proportions [26]. Several studies have clearly demonstrated that improved glycemic control is strongly associated with decreased development and/or progression of diabetic complications in diabetes. Generally, hemoglobin (Hb)A1C and is the result of an irreversible non-enzymatic glycation of the beta chain of Hb A [27 - 28]. HbA1C is used routinely to assess long term glycemic control in patients with DM [28]. The main limitation of HbA1C measurement is Hb disorder [28].

It is noted that measurements of HbA1c by HPLC and electrophoresis are clearly unsuitable for homozygous hemoglobinopathies; for heterozygous hemoglobinopathies and Hb synthesis variants [29]. Hb Cranston results from an aberration in Î² globin gene [30]. The frameshift mutation is the result leading to an abnormal elongation of the Î² chain by amino acids- (144) Lys-Ser-Ile-Thr-Lys-Leu-Ala-Phe-Leu-Leu-Ser-Asn-Phe-(157)Tyr- COOH [30]. This variant has firstly been described in USA [30]. The clinical significance of this specific unstable hemoglobin is the relationship with a compensated hemolytic state owing to an unstable hemoglobin variant [30]. The main pathogenesis is believed to due to the nature of this abnormal hemoglobin, resulting from the elongation.

The correlation between Hb Cranston and diabetes has never been proposed. The possible role of Hb Cranston instability in altered metabolic change relative to poor diabetic control has not been entirely resolved and needs additional study for clarification. In this work, the author calculated for the required energy for HbA1C per unit in Hb Cranston disorder and found that the required energy is different from normal Hb and might have some effects on the pathophysiology process.

2. Materials and Methods

2.1. Pathways for Glycosylation Reaction

Concerning the glycosylation in formation of Hb A1C, the reaction occurs between N terminal of protein and glucose. This pathway is a type of "energy consuming reaction". The required energy for complex formation is equal to 70 kCal/reaction of a Hb molecule [31].

2.2. Calculation for Required Energy of Hb Cranston Disorders

Basically, the normal Hb has the molecular weight 64,500 gram/mole. Hb Cranston is a beta Hb disorder which excessive quantity of beta globin chain [31]. Generally, the chain imbalance is the key factor initiating the damage to the red blood cell and it is the major pathophysiological event in all forms of the thalassemia [31]. Several forms of Hb Cranston are documented in hematology (Table 2). The molecular weight of mutated Hb in each category of Hb Cranston disorder was calculated. In addition, the required energy for complex formation in each category was calculated.

3. Results

The calculated molecular weight and required energy for complex formation for each category of Hb Cranston disorder is calculated and presented in Table 2.

Table 2. The molecular weight and the required energy for HbA1C formation in normal and Hb Cranston disorders

Category	Molecular weight : MW	Require energy per unit: RE (Cal)
Beta and Beta	64500	1.085
Beta and Cranston	68000	1.029
Cranston and Cranston	71500	0.979

*There are two beta globin chain in a hemoglobin molecule. Each beta globin subunit has a molecular weight of 15,867 Daltons and each Cranston globin submit has a molecular weight of 19,367 Daltons; these data was used for calculation of the molecular weight in each type of hemoglobin CS disorder.

4. Discussion

A1C is a specific nonenzymatic glycated product of the Hb beta-chain at the valine terminal residue [32]. It is normally present in circulating erythrocytes because of the glycosylation reaction between Hb and circulating glucose [32 - 33]. Recently, the nature of energy consuming reaction in formation of HbA1C was reported and this was proposed as an important underlying pathophysiology for poor nutritional status, muscle loss, and functional impairment poor control diabetic cases [31]. The correlation between the hemoglobinopathies and diabetes has been proposed for a few recent years. Although there are many studies on beta thalassemia [34] the knowledge on Hb Cranston is limited. Here, the author focuses on the energy change in formation of HbA1C in case of Hb Cranston disorder, a common hemoglobin disorder in many tropical countries.

In this work, the author determined the required energy per unit of the HbA1C formation reaction in Hb Cranston disorder. According to this study, the required energy range for any type of Hb Cranston disorder is usually lesser than that of normal Hb. This pattern is different from the previous report on beta thalassemia [10]. Although the primary aberration in beta thalassemia and Hb Cranston disorder are similarly on beta globin chain, the effect of elongation in Hb Cranston disorder is overall different from shortening of beta globin chain in beta thalassemia. Hence, it might show that the Hb Cranston disorder did significantly decrease the complication owing to energy consumption during HbA1C formation in poor diabetes control cases. The present work implies that Hb Cranston disorder might be a protective factor for complication of diabetes.

References

[1]   Lippi G, Guidi GC, Mattiuzzi C, Plebani M. Preanalytical variability: the dark side of the moon in laboratory testing. *Clin. Chem. Lab. Med.* 2006;44:358-65.

[2]   Bonini P, Plebani M, Ceriotti F, Rubboli F. Errors in laboratory medicine. *Clin. Chem.* 2002;48:691-8.

[3]   Wiwanitkit V. Types and frequency of preanalytical mistakes in the first Thai ISO 9002:1994 certified clinical laboratory, a 6 - month monitoring. *BMC Clin. Pathol.* 2001;1:5.

[4]   Maresh M. Screening for gestational diabetes mellitus. *Semin. Fetal. Neonatal. Med.* 2005;10:317-23.

[5]   Abe M, Ikeda Y, Yaginuma M, Shimizu N. Pancreatic function tests. *Nippon Rinsho.* 1969;27:291-300.

[6]   Hintz R. Laboratory diagnostic examinations in diabetes mellitus. *Przegl. Dermatol.* 1972;59:409-12.

[7]   Marks V. Biochemistry in clinical practice. Heinemann: London, 2003

[8]   Higgins C. Diagnosing diabetes: blood glucose and the role of the laboratory. *Br. J. Nurs.* 2001;10:230-6.

[9]   Petersen PH, Ricos C, Stockl D, Libeer JC, Baadenhuijsen H, Fraser C, Thienpont L. Proposed guidelines for the internal quality control of analytical results in the medical laboratory. *Eur. J. Clin. Chem. Clin. Biochem.* 1996;34:983-99.

[10]   Wiwanitkit V. Case of sudden death in venipuncture clinic. *Phlebology.* 2004; 19: 193

[11]   Howanitz PJ.Errors in laboratory medicine: practical lessons to improve patient safety. *Arch. Pathol. Lab. Med.* 2005;129:1252-61.

[12]   Wiwanitkit V. Interesting cases in aberrant biochemistry laboratory results. *Chula Med. J* 2001; 46: 997- 1001.

[13]   Valenstein PN, Sirota RL. Identification errors in pathology and laboratory medicine. *Clin. Lab. Med.* 2004;24:979-96.

[14] Wiwanitkit V. A knowledge survey of medical students about rational tube preparation. *Chula Med. J.* 2001; 44 : 349-354.

[15] Teerakanchana T, Vanichtantikul P, Manorom W. Blood glucose : collection and transportation of specimen. *J. Rajavithi Hosp.* 1994; 3: 65-69.

[16] Dietel H, Bielfeldt H. Obstetrical aspects of diabetes in pregnancy. *Geburtshilfe Frauenheilkd.* 1968;28:513-22.

[17] Wood WG. The preanalytical phase--can the requirements of the DIN-EN-ISO 15189 be met practically for all laboratories? A view of the "German situation". *Clin. Lab.* 2005;51:665-71.

[18] Maresh M. Screening for gestational diabetes mellitus. *Semin. Fetal. Neonatal. Med.* 2005 Aug;10(4):317-23.

[19] Abe M, Ikeda Y, Yaginuma M, Shimizu N. Pancreatic function tests. *Nippon Rinsho.* 1969 Feb;27(2):291-300.

[20] Hintz R. Laboratory diagnostic examinations in diabetes mellitus. *Przegl Dermatol.* 1972 May-Jun;59(3):409-12.

[21] Marks V. Biochemistry in clinical practice. Heinemann: London, 2003

[22] Wiwanitkit V. Interesting cases in aberrant biochemistry laboratory results. *Chula Med. J.* 2001 Dec; 46 (12): 997- 1001.

[23] Types and frequency of preanalytical mistakes in the first Thai ISO 9002:1994 certified clinical laboratory, a 6 - month monitoring. *BMC Clin. Pathol.* 2001;1(1):5.

[24] Petersen PH, Ricos C, Stockl D, Libeer JC, Baadenhuijsen H, Fraser C, Thienpont L. Proposed guidelines for the internal quality control of analytical results in the medical laboratory. *Eur. J. Clin. Chem. Clin. Biochem.* 1996 Dec;34(12):983-99.

[25] Lefebvre P. Diabetes yesterday, today and tomorrow. The action of the Internation Diabetes Federation. *Rev. Med.Liege.* 2005;60:273-7.

[26] LeRoith D, Smith DO. Monitoring glycemic control: the cornerstone of diabetes care. *Clin. Ther.* 2005;27:1489-99.

[27] Noda M, Izumi K. Laboratory markers for glycemia and their target. *Nippon Rinsho.* 2002;60 Suppl 9:667-74.

[28] Lahousen T, Roller RE, Lipp RW, Schnedl WJ. Determination of glycated hemoglobins (Hb A1c). *Wien Klin Wochenschr.* 2002;114:301-5.

[29] Camargo JL, Gross JL. Glycohemoglobin (GHb): clinical and analytical aspects. *Arq. Bras. Endocrinol. Metabol.* 2004;48:451-63.

[30] Weykamp CW, Penders TJ, Muskiet FA, van der Slik W. Influence of hemoglobin variants and derivatives on glycohemoglobin determinations, as investigated by 102 laboratories using 16 methods. *Clin. Chem.* 1993;39:1717-23.

[31] Bunn HF, Schmidt GJ, Haney DN, Dluhy RG. Hemoglobin Cranston, an unstable variant having an elongated $\hat{I}^2$ chain due to nonhomologous crossover between two normal $\hat{I}^2$ chain genes. *Proc. Natl. Acad. Sci. USA.* 1975; 72: 3609-3613.

[32] Wiwanitkit V. Energy consumption for the formation of hemoglobin A1C; a reappraisal and implication on the poor control diabetes mellitus patients. *J. Diabetes Complications.* 2006 Nov-Dec;20(6):384-6.

[33] Kawahara R. Glycohemoglobin (HbA1c, HbA1). *Nippon Rinsho.* 1998;56 Suppl 3:55-8.

[34] Wiwanitkit V. Beta thalassemia and energy consumption in hemoglobin A1C formation: a model. *J. Diabetes Complications.* 2007 Sep-Oct;21(5):338-40.

Index

D

F

G

H